D0395092

HUMAN EFFLORESCENCE

A STUDY IN MAN'S EVOLUTIONARY AND HISTORICAL DEVELOPMENT

HUMAN EFFLORESCENCE

A STUDY IN MAN'S EVOLUTIONARY AND HISTORICAL DEVELOPMENT

By

ELEANOR B. MORRIS WU
Toronto, Canada

WARREN H. GREEN, INC.
St. Louis, Missouri, U.S.A.

Published by

WARREN H. GREEN, INC.
8356 Olive Boulevard
St. Louis, Missouri 63132, U.S.A.

ISBN Number: 87527-323-8

Printed in the United States of America

ACKNOWLEDGMENTS

No book can be written without the assistance of many people. This book is no exception. While it is impossible to name all those who have given assistance, there are some who deserve special mention and thanks:

MR. JOHN TUNG
I-Feng Co., Hong Kong

MR. DAVID YEN
Sino-American Silicon Products, Inc., Taiwan

MR. ERNEST WONG
Rotating Memory Systems, Inc., Milpitas, California

DR. CHANG CHI-YUN
Founder, University of Chinese Culture, Taiwan

PROF. WANG KWAN-CHING
University of Chinese Culture, Taiwan

DR. LI TCHONG-KOEI
President, Pacific Cultural Foundation, Taiwan

PROF. CHEN LI-FU
Taiwan

PROF. WONG TON HAM
National Taiwan University

Eleanor B. Morris Wu

CONTENTS

PART II
"AN ACCIDENT OF HISTORY: THE ONTOLOGY OF HISTORICAL CIVILIZATION IN THE FOUNDATIONS OF THE EGYPTIAN EMPIRE"

Chapter *page*

PART III
"AN ACCIDENT OF LANGUAGE: THE SPECIFIC DEVELOPMENT OF CHINESE CIVILIZATION IN THE FOUNDATIONS OF THE EGYPTIAN 'MASTER CODE' OR *I-CHING*"

Chapter *page*

HUMAN EFFLORESCENCE

A STUDY IN MAN'S EVOLUTIONARY AND HISTORICAL DEVELOPMENT

PART I

"AN ACCIDENT OF EVOLUTION: MAN'S PSYCHIC DEVELOPMENT FROM FIRST TO FOURTH EFFLORESCENCE"

Chapter I

THE UPRIGHT-STANDING HUMANOID: A CHANGE IN THE THEATRE OF EVOLUTIONARY DEVELOPMENT FROM PHYSIOLOGY TO BIOCHEMISTRY

That the history of mankind had more than a single point of origin, that his development, evolution or cystallization into his present form was in fact more of an efflorescence rather than a single point on the evolutionary time scale, cannot really any longer be seriously doubted. The evidence of human sites as old as half a million years all over the world which are replete with evidence of fire as a technological and social basis for human collectivities, carefully sculpted stone tools and evidence of quasi-systematic burial practices at this early date, defies the imagination in attempting to postulate or imagine a single point or site of origin for man's development. This is not to say that once the efflorescence of human development occurred it also had a single linear and always progressive development, for in fact it did not. To be more precise, it can be said that mankind's efflorescence on this earth occurred concurrently with the efflorescence of mammalian life, from whom, as all scientific evidence concurs, mankind was a specific outgrowth.

The difficulty, in fact, of drawing distinct lines between mammalian efflorescence and human development can be seen from the evidence that abounds in Africa on the so-called 'missing-link' of the *Austrolopithecus*, and, further, the finds that demonstrate that varying differentiated species of mankind such as *Austrolopithecus, Neanderthal* and *Cro-Magnon* man existed in the distant past concurrently in time and in space. As well, the recent work done in Africa and other parts of the world by Godall and others on the social life of apes and other differentiated groups of the ape species such as chimpanzees, orangutans and so forth demonstrate the increasing difficulty that modern thinkers are having in defining exactly what is the *je ne sais quoi* in structure and function which differentiates man from at least certain of his mammalian cousins.

In terms of the conventional categories of physical anthropology there are few leads, since brain size among the earlier known species of mankind do not differ significantly enough for sheer quantity of cranial capacity to be a definitive marker. The same is true for cranial capacity of many apes in comparison with *Homo Sapiens*, and indeed some mammals such as the elephant and the whale would seem to have even larger cranial capacity than *Homo Sapiens* or any simian species. We are left with the conventional wisdom that mankind's upright stance, so similar to the stance of many ape species, as well as the minute differentiation of highly specialized use of hand and thumb, are the critical physiological features which differentiates *Homo Sapiens* from his mammalian cousins, a feature which is surely only a difference of degree, of quantity, rather than quality.

But, of course, it is not the physiological differences between man and apes which are the real hallmarks of mankind's differentiation from the animals, but rather the cultural and technological aspects of *Homo Sapiens* as a collective species. That is to say, if tomorrow a group of apes who differ so slightly from present day mankind physiologically, were to acquire the trappings of culture and technology, we would have to term them as part of the collective species of *Homo Sapiens*, and given the enormous time span in which such judgements must be made it is entirely possible that at some point in the distant future, should these present day ape collectives survive, this will happen. The question of course is whether present day mankind by the time this has happened would be the same as it is today, or whether it would have achieved some new stage of physiological and/or cultural development that would make mankind at that stage as different from the newly evolved apes as mankind is today as different from present day apes. Of course this is the kind of hypothetical question that opens all kinds of logical conundrums and caveats, the kind in fact that have made it so difficult for modern day thinkers to develop a plausible picture of *Homo Sapien* development in the first place.

It must be remembered that mankind as far as we know has existed on this earth already for one to two million years. During this time there is ample evidence to support the claim that *Homo Sapien* development was anything but linear and that in many periods *Homo Sapiens* existed side by side with more 'ape-like'

Homo Sapien sub-species such as *Neanderthal* and possibly *Austrolopithecus.* Prior to this 1-2 million year threshold it is even more likely that mankind in a pre-efflorescent state existed and existed side-by-side with a larger profusion of 'man-like' apes or 'ape-like' men who no doubt existed in a great variety of sub-specific forms. The present existence, in fact, of apes today might well be the lost vestiges of the evolutionary adaptation of man as a species, man being proven the most adaptive of the sub-specific forms of mammalian development that occurred prior to 1-2 million years ago, leading eventually by 100,000 years ago, in the upper Pleistocene, to the virtual extinction of other sub-specific 'man-like' forms who proved less adaptive to the emergent *Homo Sapiens.* It is exactly this kind of thesis which has been explored so well, for lack of a more conventional forum, in such films as "The Planet of the Apes" series.

The thesis so brilliantly developed and expostulated in these films is that human development is an ongoing saga of competition among similar but not identical ape-like forms, which in the enormous span of time in which such competition is given reign to play itself out, leads to bizarre variegations and enormous vicissitudes in the historical experience of these competing forms. Very adroitly it is illustrated that once a given ape-like form has reached a certain physiological threshold *sans* culture or technology, a simple adaptive mutation in even a single individual of that species may, by virtue of the simian proclivity for imitative behavior, almost, as it were, suddenly lift an entire simianoid collectivity beyond a critical threshold to apprehension of culture and technology, so that the collectivity might be thrust, as it were, overnight from a simianoid to a humanoid mode of behaviour.

Conversely, it is illustrated in these film series how, given certain extreme conditions of repression and oppression by one humanoid species over another, regressive imitative behaviour may take hold of an entire collectivity, so that bit by bit repression would lead to what would appear to be an almost sudden reversion of humanoid behaviour to simianoid. Loss of culture could become so extreme among a humanoid collectivity, including loss of language or verbal communicative ability, that the collectivity would sink down beyond the critical threshold of humanoid behaviour once again so that it would no longer, by present day standards, be classified as *Homo Sapien*. The point here is that cultural and technological acquisition by a humanoid collectivity is almost a matter of accident rather than necessity, and accident as opposed to necessity may have many vicissitudes. The accoutrements of culture and technology are not inborn in *Homo Sapiens*, although the potentiality for such accoutrements is. The single mutation which would allow for this potentiality to become actuality is not definitive enough to preclude any humanoid collectivity from either happening to acquire culture and technology or happening to lose it.

The nature of the simple mutation which differentiated the critical threshold of culture and technology from the lack of it, or humanoid from simianoid behaviour, was obviously not of a gross physiological nature, since absolute cranial capacity has not been found to be a definitive marker between man and ape or among the variegated sub-species, the remains of which have been found in archaeological sites and which lie in that grey area between simianoid and humanoid forms. The nature of this simple mutation would have to be, therefore, neurophysiological, that is, concerned with the infinitely complicated neural structure of the brain, which is beyond the means of present day science to analyze properly even in extant species, or to put even more simply, biochemical. In order to target the likely nature of such biochemical mutations it is necessary to first look at the gross physiological features which differentiate the mammalian species from the reptiles and other species which precede mammalian genuses in evolutionary time.

The two gross features which most markedly differentiate mammalian species from their predecessors are warm bloodedness, the inter-uterine bearing of the young by the female of the mammalian species, and subsequent suckling of the young upon birth by the female. That which is most blatantly new in terms of evolutionary time with the appearance of the mammalian physiological type is the high degree of sexual differentiation among mammalian species and in particular the high degree of exotic differentiation of the female of the species from the male. While among all living creatures above single-celled animals there is a biochemical differentiation between sperm of the male and ova of the female which provides for reproduction of the species, among vertebrates, at least, there is little gross physiological distinction between the sexes until the stage of mammalian evolution has been reached. The deposition of embryo in egg form is more or less randomly distributed among various pre-mammalian species according to sex, and the task of nurturing these eggs until the time of their hatching into almost full-formed individuals of the species may also be more or less randomly distributed among the various pre-mammalian species between the male and female of the species. Gross physiological differences between male and female in pre-mammalian vertebrates, where it exists, is of a superficial rather than morphological, nature.

It stands to reason therefore that the nature of the differentiating markers among various mammalian species is related to a more highly differentiated difference between the male and female of these species. However, since the gross physiological features that distinguish male from female among mammals remains an absolute criteria common to all mammalian species, the kind of differentiation that might be seen to occur between various kinds of mammalians would be biochemical rather than physiological.

Recent research in biochemistry has shown that neurotransmitters or the chemical messengers which the brain uses to 'think' are remarkably similar, in a chemical sense, to female hormones. Thus norepenephrine which is the chemical messenger used in the cerebellum or the 'upper brain' or the brain cavity is extremely similar to the female hormone estrone and certain of its derivatives, while the neurotransmitter or chemical messenger of the medulla or 'lower brain' or autonomic nervous system, while chemically similar to both female and male hormones, is perhaps more similar to the major male hormone, testosterone. It might be said even from a gross physiological point of view that while pre-mammalian vertebrate species possess sexual differentiation, there exists no defined 'female type' until the stage of mammalian evolution is reached. From an adaptive evolutionary point of view, in fact, it can be said that the development of a distinct 'female type' is an evolutionary luxury that was not reached until the stage of mammalian evolution. Struggle for survival in pre-mammalian forms was of such an acute nature that the fine sentiments or the loss of the individual's sense of personal survival, even for a short period of time, which are the necessary evolutionary trappings of the 'female type' known to mammalian species, was not possible and was maladaptive.

In pre-mammalian vertebrates, the deposition of eggs and their fertilization in no way affected behavioural modes which might have interfered with the individual's ability to cope with environment in the pursuit of individual survival. Cold-bloodedness as a trait of pre-mammalian vertebrates likewise suggests that ability of the individual to respond immediately to the environment, changing the temperature of its blood with the temperature of the environment. Thus in pre-mammalian vertebrates the cerebellum is small and ill-defined, the automatic or 'lower' nervous system being the main mode of neurological activity and almost entirely of a reactive rather than cognitive nature. In pre-mammalian vertebrates it can be said that aggression and reactivity are the main modes of behaviour, where the individual either forces the environment to bend to its will or else, when this is not possible, automatically bends itself to the will of the environment, all in the quest for individual and not specific survival.

This is not meant to be a value judgement, to assert that the 'male type' corresponds to a more primitive form of animal evolution, while the 'female type' corresponds to a more advanced one. Nor is it meant to assert that the 'male type' among mammalian species is rooted in the lower nervous system of the medulla, while the 'female type' is rooted in the higher nervous system and cognitive functions of the cerebellum. The truth is that no animal species can afford to bypass aggressive and reactive behaviour to the environment if it has any hope of surviving, reproducing and developing, and had the mammalian species lost this mode of behaviour it would quickly have become extinct.

Nonetheless, the conclusion is inescapable that in mammalian species at least part of the modes of aggressive and reactive behaviour have been abdicated, and where this abdication evolved was in the development of a more highly differentiated female or 'female type.' While such abdication is not so evident in lower mammalian species where the terms of gestation are short, the litter often large, and the physiological features of the female relatively unchanged during periods of inter-uterine gestation or pregnancy and menses, among the higher mammals where the obverse is the case, such abdication can be seen clearly to exist. Among higher mammals where the female is periodically relatively incapacitated, there are physiological and biochemical compensations in the 'male type' of the species. Thus the male hormone testosterone and its derivates such as androgen among mammalian 'male types' produces individuals who are, relative to the 'female type' of a given species, larger in size, more muscular, and are sometimes equipped with highly specialized defensive appendages such as horns or antlers.

Nonetheless, no matter how muscular or well equipped the male type of the mammalian species may be it cannot be denied that at least periodically he is called upon to do the work of survival and adaptation for two or more, and has thus *de facto* adapted beyond the necessities of individual survival to the necessities of group or collective survival. In addition, no matter how powerful or able he may be, the 'male type' of the mammalian species is called upon to perform what is an almost impossible goal from the point of view of pre-mammalian vertebrates—the insurance of survival beyond the single individual 'self.' In the theatre of intense competition which makes up the evolutionary struggle, this is no mean feat. From this perspective, it can be seen how mammalian species, by virtue of adaptative necessity rather than instinct *per se*, were forced to develop collective or social modes of behaviour, with the major burden of aggressive and reactive behaviour falling to the males of the species as a collective rather than to the collective as a whole. As mammalian species developed more complex forms, this difference between the behavioural patterns of the males of a collective and the collective as a whole became more pronounced. No doubt this behavioural differential between the males of the collective and the collective as a whole evolved in tandum with the greater and greater abdication of aggressive and reactive patterns of male hormones, concomitant, inevitably, with longer gestation periods for the female, reduction in the size of litters, and the increase in time span necessary for mammalian young to achieve individual self-sufficiency.

Whether collective mammalian behaviour patterns resembled that of a pack where mammals lived in arctic climates and behaved as predators or hunters, or that of a herd where the species lived in temperate climates and behaved as grazers, or that of clusters where mammals lived in tropical climates and behaved as gatherers, these basic behavioural patterns persisted. The micro-variations that occurred among various mammalian species were adaptive in terms of the kind of climate and ecological niche they inhabited. Thus, in colder climates where the source of food was scarce, mammalian species evolved pelts, fur or thick matted hair and often small or medium body size to allow for minimal food intake and facile body temperature regulation. In temperate climates, evolving herds developed thick hide with minimal hair to protect against the vagaries of temperature in those climes and as protection against the ravages of predatory insects. In tropical climes the clusters of mammals, for reasons of extreme external heat, lost a percentage of body hair without developing thick hides, while the body hair that did remain was long and thinly matted. While in all cases these micro-variations among mammalian species took tens of millions of years to become defined, as different mammalian collectives found themselves isolated into rigid ecological niches and temperature enclaves, these basic behavioural patterns common to all mammalian species persisted.

While the basic upright stance common to simianoid mammals is commonly attributed to the need of simianoids to escape from predators in trees, developing long arms to enhance their quasi-arboreal modalities, it is by no means certain that for the duration of the evolutionary process during which time simianoids developed and were eventually evolved into *Homo Sapiens*, that they were confined to tropical climes. The present day apes who inhabit tropical climes could as well be specific differentiations of a proto-simianoid species, or even a regressive form of that species, as the original ancestors of *Homo Sapiens* from whom the latter evolved in a continous and unbroken line. In fact, it is quite probable that a proto-simian species developed first in tropical climes, and then, due to its greater adaptative ability, proliferated throughout all given climes, and in the process accomodated itself to various social modes of behaviour which were common to earlier mammalian species who lived in these various climes. That is to say, the proto-simian mammalian species represented a critical threshold in mammalian development which enabled it to proliferate, much like the development of Cro-Magnon in the lower Pleistocene. This did not mean that the proto-simian species however was bound and chained to its clusteral form of behaviour with gathering as its main mode of food procurement. Once proto-simian proliferated into various ecological niches and temperature enclaves he would have been able to readapt social modes of behaviour, such as herd or pack, that were patterned into the responses and neurology of the early mammalian species of which he was such a close derivative.

The reason this can be inferred is that the clustering mode of behaviour from whose roots simianoids probably derived originally is no more characteristic of the humanoid derivatives of simianoids than herding or pack behavioural patterns. If anything, the small, closely knit groups characteristic of simian behaviour in clusterings, where the simians are neither predators nor habitually carnivorous, would seem to represent one of those chance mutations in animal development that changes the course of evolutionary history, but which is highly particular and not within the framework of general species development. In other words, while the particular ecological niche of the original simian may have produced the right ecological conditions in which more or less upright stance became an evolutionary advantage, it does not mean that once this mutation was achieved the evolved species would have been confined, by reasons of such mutation, to that particular ecological niche and temperature enclave. On the contrary, while more or less upright stance was developed as an adaptation to arboreal modality, once this mutation was achieved, it had adaptive advantages far beyond that particular ecological niche and temperature enclave. Whereas four-footed animals had only their mouths and teeth to use as foraging tools, the semi-upright simianoid had two hands as well as a mouth and teeth to use as foraging tools and even weapons, making him highly adaptive as both a herding and pack animal.

Another powerful reason for inferring that proto-simians proliferated and were not confined to a clustering form of social behaviour and a gathering mode of food procurement, is the nature of the biochemical and bio-psychological evolution among simianoids which eventually led to the development of humanoids. It is extremely unlikely that if humanoid development was in a continous and unbroken line from simianoids similar to present day apes with their particular modes of social behaviour and food-gathering habits that humanoids, both prehistoric and present day, would be imbued with the characteristics that they did and do indeed possess. Archaeological sites of early humanoids or late simianoids demonstrate conclusively that these species were carnivorous, lived in all kinds of climates and ecological niches and were not confined to small, individual clusters of individuals. All evidence indicates, from the remains of slain animals buried closely with the remains of late simianoids or early humanoids in the lower Pleistocene 1-2 million years ago that these species were both aggressive and carnivorous, and that their collective behaviour was not confined to small, individual clusters. This evidence is in keeping with the general, rather than any particular, trend of mammalian development.

If mammalian development had changed its trend at this critical threshold of proto-simian development and had remained in small, individual clusters as a mode of collective behaviour, the elaborate social patterns of early mammalian development which allowed for the proliferation of species would have been inhibited. The larger the collectivity of males in a species, the easier for any single male to carry out his evolutionary mandate to protect both himself and the female of the species at the same time. In the savage arena of evolutionary struggle, it was an onerous task to provide for his own individual survival, not to mention carrying the burden for another who was periodically helpless with gestation and the nurturing of young. The clustering of present day simians in their protected tropical jungle niches is often not much larger than a single extended family. In such a protected niche, the struggle for survival is not so arduous as in less favourable niches and climes, and groups of males may remain small without being at an evolutionary disadvantage. For proto-simianoids to have proliferated as he so clearly did, however, he had to face inhospitable niches and climes where larger collective groupings would have been a necessity.

Similarly the lack of predatory and carnivorous modes, while not evolutionarily maladaptive in protective niches and climes, would have been disasterous in less favourable ones. In favourable niches and climes, the aggressive and reactive instincts of the 'male type' common to late simianoid and humanoid species could necessarily be muted, but in the actual theatre of evolutionary struggle, beyond one particular, privileged and necessarily limited kind of niche, it would by necessity become pronounced. Even in the protected niches, over time, climate changes, jungle becomes forest, forest becomes plain, plain becomes steppe, and steppe becomes snow bound as glaciers move and recede over the surface of the earth in the evolutionary time span. While it was very possibly in the protected jungle niche that proto-simianoid, as it were, by accident, developed his semi-upright stance, it was this accident or mutation that allowed mammalian evolution to progress, succeed and adapt more successfully than ever before, in all kinds of climes and evolutionary niches, because this mutation, which occurred under such particular circumstances, in fact made proto-simian such an exquisitely adaptive animal.

With the onset of the semi-upright stance of proto-simian which in no way inured either the aggressive instincts or the social needs of the 'male type' of the species, there may have occurred complications which were tantamount to evolution in the 'female type' of the species. Gestation and childbirth became more cumbersome and difficult as the weight of the fetus and the expulsion of the newborn was far more difficult on two legs than it had been on four. Over time, it would become evolutionarily adaptive in terms of the survival and fitness of the 'female type' for the litter to be reduced more and more frequently to a single offspring for whom the lactation, previously available to many newborn, would be entirely available to a single offspring, making the period of its suckling longer and consequently prolonging both the period of helplessness of both mother and offspring. As the periods of suckling were prolonged, so were the periods during which profusions of female hormones in the system of the mother were prolonged, the periods in which normal productions of androgens and testosterones in the 'female type' would be shortened. Eventually such long periods of female hormone production would lead to longer periods of gestation for the fetus, lengthening again the period during which both mother and offspring were useless in the necessary adaptive struggle for individual survival. Thus began the evolutionary process which ultimately produced the 'female type' of late simianoid, early humanoid development. Concomitant with this process, the 'male type' in his mode of collective male behaviour would, by necessity, in order to allow him to fulfill his instinct for preservation and propogation of the species, become more pronounced as well.

While in every human male there exists both an 'x' or female chromosome and a 'y' or male chromosome, and in every female there exists only two 'x' or female chromosomes, evolutionary adaptation along the lines suggested would have allowed the male to fulfill the potential of his 'x' or female chromosome as well as his 'y' or male chromosome. As periods of gestation and suckling become prolonged, the time in which the male fetus and infant and even child remained helpless and docile, deprived of the potential of his 'y' or male chromosome, also became more prolonged. This allowed time for the humanoid male to become infused with the female hormones so necessary to his cognitive development, delaying the time at which the male hormones necessary to his aggressive and reactive instincts were produced in profusion in his system. As well as increasing his cognitive abilities, the biochemical consequences of long periods of gestation and suckling would enhance his capacity for social and collective behaviour. For social and collective behaviour to be successful among a group of individuals, each individual must have the ability to suppress or mitigate to a certain extent his own primal instinct for individual survival, replacing at least part of this individual survival sense with a collective one. Similarly individual male aggression must be replaced at least in part with a sense of collective male aggression. While this form of behaviour is evident among almost all mammals to varying degrees, nowhere is it more pronounced, of course, than among *Homo Sapiens*, and there is every reason to believe that this was one of the behavioural markers differentiating man's early ancestors from the other mammals among whom he lived and preyed upon.

It may be said, then, from existing archeological data, that the first efflorescence of late simianoid or early humanoid efflorescence occurred with the advent of an *Australopithecus* type some 1-2 million years ago. The next efflorescence where evidence of Neanderthal and Cro-Magnon man is found is 500,000 years ago, and this second efflorescence represents a more advanced type of late simianoid or early humanoid type, as the skulls and fossil remains in many cases are really indistinguishable from skeletal remains of present day *Homo Sapiens*. In addition, while *Australopithecus'* sites reveal no evidence of set cultural or technological patterns, no evidence of communal burial and no use of fire, all of these things are in evidence by the time of the second efflorescence of the humanoid type some 500,000 years ago in the lower Pleistocene. However in sites of *Australopithecus* there is evidence of communal life, that is, *Australopithecus* travelled in groups and returned to the same locations time and time again, where they dwelt at least periodically for long periods of time as the vicissitudes of environment and climate permitted. Although these ecological patterns in habitats of *Australopithecus* in fact differ but little from the patterns of almost all higher mammalian forms, nonetheless it is necessary to postulate that *Australopithecus* with its semi-upright stance marks the onset of the efflorescence of the late simianoid or early humanoid type everywhere in the world, an onset which in the course of a 1.5 million year time span led eventually to a new threshold of humanoid development represented by the Neanderthal or Cro-Magnon man of the lower Pleistocene 500,000 years ago.

While 1.5 million years may seem to be a span of time that defies the imagination of human beings living today who are aware of a historical period of their own species no longer at the very most than 15,000 years, it must be remembered that in terms of the evolutionary development of life on this planet even 1.5 million years is only a brief moment in time. As mammalian life began on this planet 63 million years ago, it took about 62 million years for mammalian life to achieve the critical threshold of late simianoid or early humanoid development. It is thus important to remember that mammalian life forms prior to the development to the first efflorescence of late simianoid/early humanoid forms must have, as well, passed through innumerable cycles of efflorescence, and further adaptation leading to further cycles of efflorescence similar to the cycle we have described for the evolution of an 'accidentally' differentiated upright-standing simianoid. It must also be noted that with the efflorescence of humanoid type, mammalian evolution as a whole did not suddenly cease, and the mammalian life that exists on this planet today, as has already been suggested, represents well established types of successful environmental and climactic adaptations of mammalian forms. The variety of such types that presently exist represent the natural evolution of successful adaptive types to particular environment and climates, all of which continue to contend for future efflorescence of the most successful types in future evolutionary time.

In future time, this will lead to the eventual formation and efflorescence of further successful 'proto' mammalian types. Thus in the 63 million years in which mammalian life has existed on this planet it is logical to assume that there have been innumerable proto-types which were in efflorescence throughout the world prior to the development of humanoid types, the latest but by no means for certain the last attempt that mammalian life will continue to make in establishing more successful adaptive types on this planet. One of the points to be made from this perspective is that for mammalian life, having firmly established itself as a four-footed creature in the course of 63 million years, the development of a two-footed mammal as the dominant type of that genus is, at least in terms of the time element involved, still an experimental species development. It is thus not without sound scientific fact that present day mankind does and should continue to question the fate of his species' survival on this planet. There is no doubt that certainly at least since the upper Pleistocene 100,000 years ago, the efflorescence of *Homo Sapien* on this planet has begotten a radical change in the quality of life forms among mammalian species. But the radical biochemical and neurological changes in the upright-standing mammal whose onset was 1-2 million years ago, leading in turn to complex psychological and behavioural modes derived from these biochemical and neurological changes, has not, from an evolutionary time perspective, necessarily had the time to develop the kind of homeostasis or equilibrium that 63 million years of evolution among four-footed mammals permitted.

To put it another way, collective and behavioural modes among four-footed mammals are stable in a way that no one living today could claim to be true for two-footed, upright-standing mammals. While stability of collective and behavioural modes is in and of itself not necessarily always an adaptive advantage, it does suggest at least that as far as the humanoid type is concerned, the strategy of evolutionary survival may be far from over. Nonetheless, in looking at the future in the same way as the past, while we may envision that further differentiations of the humanoid type might occur, leading to an efflorescence of that future type, it is absolutely impossible to know of what this differentiation may consist. While conceivably it could be of a gross physiological nature, it could as easily be of a biochemical nature only, leading to new forms of cultural adaptation, or even simply one new form of social and cultural adaptation, thus ensuring the establishment of the humanoid type on this planet as dominant in its way as the mammalian type that preceded it, or the reptilian type which in its time preceded the mammalian. If it is this kind of evolutionary adaptation that is in store for mankind in the future, *Homo Sapien* would graduate, from the point of view of the categories we use, from a species development and efflorescence to an entire genus of development and efflorescence.

The important point to remember here is that since the appearance of the earliest humanoid type 1-2 million years ago, mankind's evolution has been of a biochemical, neurological, psychological, cultural and social nature, rather than of a gross physiological nature. It is in this respect that the pattern of humanoid evolution differs so markedly from the pattern of evolution of all species and genuses that preceded it on this planet. While in the evolutionary pattern of species and genuses prior to the onset of humanoids, gross physiology was labile in evolutionary time, this would not seem to be the case for mankind, for his physical form is more or less fixed. The physiological markers which distinguish the so-called 'races' of mankind do not represent any differentiation of gross physiology, and even in such cases which might be imagined to have existed, or exist in the future, such as illustrated in the model presented by the 'Planet of the Apes,' where differentiation among different races, groups or families of humanoids occurs, this differentiation is, at most, of a minor and labile biochemical, or even simply, cultural and social nature, rather than physiological as such.

A study of this kind of differentiation belongs properly in the realm of humanoid evolution as it is perceived to develop in realms that are entirely of a psychological, cultural, social and political nature. An examination of the psychological, cultural, social and political evolution of mankind is the proper study of mankind's evolution and differentiation, if we are to adopt the plausible argument, towards which all evidence available to us points, that mankind as a type represents a new stage in the evolutionary process, not merely of mammalian life, but of all life, and must be considered, for all practical purposes, even at this early point in time, as a genus rather than a species.

Let us remember that the major markers which differentiates the mammalian genus from the genus that preceded it are warm-bloodedness and the inter-uterine gestation and suckling of the young by the females of the mammalian genus. The complex systemic hormones which accompanied the gross physiological changes from the reptilian to the mammalian genus underwent radical transformations and further complex changes with the onset of the first humanoids. In reptilian species, and genuses which preceded them, reactions to the environment and climate were often of a gross physiological nature. In reptiles, blood temperature changes with the temperature of the exterior world, color of the skin or shell may change with environment, and even organ regeneration is also not unknown. In the mammalian species, however, gross physiology does not change in reaction to the environment; rather, hormonal systemics change in reactions to the environment, and in upright-standing mammals or humanoids, hormonal systemic changes in reaction to the environment lead to tangible but labile patterns of psychic and cultural systemics, which is, of course, one aspect of the relative instability of mankind's behavioural and collective modes. In fact, so labile are the psychic and cultural systemics of mankind that not only are they influenced by hormonal systemics, but hormonal systemics also critically influence, in turn, psychic and cultural systemics.

Modern medicine has proven conclusively that stress of a psychic or cultural nature can influence the hormones that regulate the blood pressure as well as the regulation of sex hormones in both men and women in almost infinitely complex ways. Depending on the degree and the quality of psychic and cultural systemics of any given individual in reaction to the collective, which for humans is part of the 'environment,' hormonal activity will be influenced in ways that are either adaptive or maladaptive in terms of the very survival of the individual. The thought may kill, or to paraphrase the saying, hope may give life. The nature of the human type is such that there can be no clear line drawn between his psychic and cultural products and his hormonal or biochemical endowment, and between the two of these there is a constant feedback and interrelationship. In a sense, then, mankind possesses a biochemistry that has as it were a life of its own: hormones influence mentation, mentation influences hormones, and thus by deduction the biochemistry of mankind is a self-evolving entity. Thus, at least in this stage of humanoid evolution, gross physiology may remain as a fixed form while the theatre of evolution changes to another stage, that of biochemistry, or more precisely, neurological biochemistry. This is the radical evolutionary marker of the humanoid type that developed as a potentiality with the onset and efflorescence of *Australopithecus* 1-2 million years ago, and that developed as an actuality between the lower and upper Pleistocene 500,000-100,000 years ago.

The psychic, behavioural and collective changes that can be isolated at two thresholds, one in the lower Pleistocene 500,000 years ago and the other in the upper Pleistocene 100,000 years ago, produced the cultural and social products of *Homo Sapiens* and established new set modes of behavioural patterns, of which present day mankind is only the most recent elaboration. The first leap for *Homo Sapien* lay at the point of the lower Pleistocene where humanoids developed, seemingly in an almost sudden fashion, from a species of two-footed mammals whose behaviour was in almost no way different from other highly developed mammalian species, to a species whose behaviour would set humanoids off forever in a distinctive fashion from their mammalian brethren. From wandering collectives of grazing, herding, gathering and preying simianoids with mammalian behavioural patterns in an approximate 1.5 million year time period of duration, the humanoids of the lower Pleistocene discovered the use of fire, which became the center of their collective existence, discovered and developed methods of producing complex and specialized flint tools, and engaged in sporadic, but nonetheless deliberate, burial of their dead in communal sites.

The deliberate and continous use of fire in collective sites no doubt had implications that were psychological, cultural, social and religious for this new efflorescence of the human type. No doubt late simianoids and early humanoids had witnessed fire as an accidental occurrance on the plains, steppes and forests of their habitats during the 1.5 million years of development from *Australopithecus* to Neanderthal and Cro-Magnon of the lower Pleistocene with a certain amount of recognition of its utility and danger, as indeed mammalian species and even reptiles might do with appropriately varying degrees of clarity. But as humanoids' radical complex of hormonal and neurological biochemistry evolved and eventually stabilized at a certain threshold during this 1.5 million year time span, the mental capacity of humanoids progressed to a point where humanoids were able to appreciate the significance of a phenomena such as fire in a way in which no other species was capable. The kind of mentation that the biochemically evolved brain of humanoids was capable of differed, perhaps, from other mammals not so much in quality as in quantity. As a receptor, the untutored but fully biochemically evolved brain of humanoids resembled, in comparison to other mammalian brains, a clear picture tube, while the mammalian brain resembled a fuzzy one.

Due to the complex hormonal systemics of the biochemically evolved humanoid brain, it was capable of both spatial and temporal displacement. The apprehension of the danger produced by fire to the humanoid could be transferred in his brain to a similar situation that he was able to imagine, either as a past event or a future anticipation. If fire were dangerous and frightening to him, it might also prove frightening to his mortal enemies, those with whom he competed in his ecological niche for survival. For many hundreds of thousands of years, it was probably viewed as one of the many crude tools which humanoids did not produce themselves but used when happening upon them by chance, and which he was unlikely to differentiate from the weapons of his own person, such as his hands, arms, feet or teeth. Nonetheless, in the long process of utilizing fire for its heat and burning properties, as a weapon, humanoids were also able to recognize that its property of light was also useful as a weapon. In the night, the light of an accidental fire made it possible for man to hunt his prey and to ward off his mortal enemies. It was no doubt this particular recognition that eventually inspired humanoids to control and contain fire in his collective sites. The recognition of the useful properties of fire's light was for the humanoid his first experience of self-recognition, for the light of the fire enabled him to see where he could otherwise not, and allowed him to manipulate other creatures to see danger where they would have otherwise been unafraid and menacing. After this all-important self-recognition through the light of fire, humanoids were led then to the next inevitable step of attempting to contain and control fire in a systematic way and under the necessarily controlled conditions of collective existence.

This systematic control and containment of fire under conditions of collective existence had extraordinary implications for the further psychic and cultural development of humanoids. To systematically control and contain fire it was necessary for humanoids to learn a routine, by which this could be accomplished. Over tens of thousands of years the repetitive process by which fire could be contained and controlled began to inculcate in humanoid collectives a sense of cause and effect, a division of labor, and in fact, by the upper Pleistocene, made of the humanoid type a creature who already possessed technology.

Just as the upright stance of the late simianoids resulted in the first efflorescence of the human type 1-2 million years ago, so the acquisition of the routinized method of containing and controlling fire resulted in the second efflorescence of the human type in the lower Pleistocene about 1.5 million years later or 500,000 years ago. It was in this second efflorescence of the human type that humanoids acquired, among other things, self-consciousness and technology. The continuing perfection of the control and containment of fire in collective settings, where humanoids recognized that they could in fact 'see,' led to a plethora of other recognitions which *In Toto* amounted to a considerable sense of self-consciousness. 'Day' was recognized as being distinct from 'night,' the seasons in their turn became recognized as such, group member was distinguished from outsider, friend from enemy, man from animal, and man from woman. With the continuing perfection of the routinized process which allowed fire to be contained and controlled, the sense of cause and effect that resulted allowed man to develop other systems for the making of weapons and more specialized tools.

Among the flint tools found in the sites of the lower Pleistocene that have been identified by modern archaeologists are those which have specialized uses, for cutting hide, or polishing hide, butchering meat, polishing wood, and of course piercing or penetrating living creatures for the purposes of killing them, weapons of war and hunt. In all cases these flint tools are comprised of intricate faces and systematized edges whose full significance defies modern imagination. Only the nature of the final cutting edge can be understood, but the intricate systems that exist on the edges and faces of these tools are regular and routinized according to the particular cutting edge of the tool. In this second efflorescene of the human type in the lower Pleistocene, an intricate logic had already been developed that had no doubt psychological, social and religious implications, so complex that, lacking a key to the translation of these systematized etchings, their significance must remain forever mute to modern mankind.

In addition, the sporadic but deliberate burial of the dead in collective sites illustrates that not only had the humanoids of the lower Pleistocene gained a consciousness of the many dialectic oppositions which the control and containment of fire in and of itself had revealed to them, such as the distinction between day and night, seasons, and man and woman, and so forth, but they also had a relatively clear consciousness of their own mortality, for they were now able to distinguish between the dead and the living. The human type had discovered self, and at the same time had discovered that self was limited by death.

Undoubtedly by the time of the lower Pleistocene when the second efflorescence of the human type emerged, a rudimentary language had formed; otherwise the sense of cause and effect which was expressed in the control and containment of fire, and systematized routines which produced standardized flint tools, and the division of labor involved in both of these things, as well as the burial of the dead, would not have been possible. This does not mean that the language possessed by the humanoids of the lower Pleistocene would necessarily in any way resemble any known living language of today, although given the nature of the humanoid's gross physiology it must have been at least in part a vocal language, although it was no doubt accompanied by gesturing of any given number of body parts and signs. This is evidenced in the fact that during the next 400,000 years, during which time the humanoids of the lower Pleistocene evolved into the humanoids of the upper Pleistocene, the major development, in terms of the cultural and technological products of the human type in its third efflorescence, was the development of a form of written language. This fact is overtly evident in the picture writing in such sites of the upper Pleistocene as the caves of Lescaux in France, where written signs and abstract figures on the walls of these caves amount to a rudimentary system of written hieroglyphics, whose significance in fact remains as mute to modern mankind, lacking the proper key to its translation, as the intricate edges and faces on the flint tools of the lower Pleistocene.

In these sites of the third efflorescence of the human type, in addition to finding evidence of a written language, and regular burials and burial procedures, there are also found small carved stone figures of pregnant females. In his third efflorescence the human type had learned not only to distinguish life from death, but also birth from death. He had transcended the gnawing and stultifying sense of self as limited by death, and had achieved a transcendent view of self in a consciousness of the life-giving and creative phenomena of birth. The human type had come to recognize the unique and special role of the female in the collective, in the species and in life itself; he had discovered hope, faith, God and a sense of salvation.

It was no doubt the development of a written language, evolving slowly from the rudimentary spoken languages of the lower Pleistocene and the intricate, learned and transmitted systems of logic responsible for routinized flint tools, which enabled the human type to obtain a sense of permanence in the upper Pleistocene. In the knowledge that his painfully learned and transmitted experiences and technologies would outlast the painful sense of limited self man must have experienced in the terrible darkness of his second efflorescence, there were created the foundations of culture and society whose legacy we still enjoy in these latter stages of mankind's efflorescence. Given the fact that judged by the shortness of an individual life span, the time of evolutionary progress seems painfully slow indeed, nonetheless the development of a simple written hieroglyphic language to a more complex hieroglyphic language to present day written languages, must be regarded as in a contiguous flow of evolutionary development. Such a radical distinction as that between spoken language and written language, or no language at all and a rudimentary spoken language, are clear and distinctive enough to rate as markers of different phases or efflorescences in evolutionary development. Nonetheless and while modern day written languages and cultures differ only in degree from the 3rd efflorescence of mankind in the upper Pleistocene, it is these differences in degree as we have seen which are often the hallmark of further evolutionary adaptations.

Thus, such differences in degree rate as markers at least from the point of view of evolutionary hindsight, of a new evolutionary phase. However, by the time of the upper Pleistocene 100,000 years ago in the third efflorescence of the humanoid species or genus it is clear that the human type already possessed written language, art, religion, culture, social organization, trade and a sense of psychology, as well as the spoken language, technology and a sense of science he inherited from his predecessors, the human type in the 2nd efflorescence in the lower Pleistocene. In addition, he might well also have possessed a sense of philosophy, politics and a rudimentary form of military organization. Why modern mankind has been unable to relate himself in any cogent way to a clear line of development from the upper Pleistocene as a matter of historical experience is a worthy subject of continued investigation.

Chapter II

STABILIZATION OF THE 'FEMALE TYPE:' HUMAN EVOLUTION AND THE ASCENDANCY OF THE SUPEREGO

The statement that human life is nasty, brutish, and short would certainly be acknowledged to have been even more true in man's distant past than in modern times. Yet the concept or myth of 'natural man,' happy and unconscious in his animal existence in prehistoric times, is also a persistent notion in man's consciousness. Both notions are no doubt valid in varying degrees depending on the period scrutinized in humanoid development. There is every reason to suppose that in the case of *Australopithecus* and in the 1-1.5 million year time span of the humanoid's first efflorescence the human type differed but little in his collective and behavioural patterns from other highly developed four-footed mammalian species, for he had not yet gained consciousness of self. Once the humanoid type however gained consciousness in the lower Pleistocene 500,000 years ago, with his knowledge and use of fire, his development of specialized flint tools and his sporadic burial patterns, the phenomenological quality of life, at least by modern standards, must have been on the whole highly unpleasant.

Consciousness of self would have been an experience in violent opposition to the natural environment in which humanoids lived and were almost entirely a part. The kind of bizarre contradiction that this suggests for the quality of man's life in these times is very much akin to the experience of a small child, two or three years old, who senses selfhood without the means of expressing or fulfilling it, and who has no parental guides to care for him and provide for his needs during this most frustrating time of human development and growth. Yet, the humanoid collectives of this time were not confined to small children; they were composed of physically and emotionally ripe adults with this kind of infantile mentality, and the brutality that they must have been capable of practising on each other and themselves in this untutored stage of collective development defies the imagination of modern mankind. The sense of self as it emerged in fully physical and emotional ripeness resembled nothing so much, at least by modern standards, as a violent illness, or an unquenchable melancholy, the constant *angst* of which could only be relieved by violent individual acts and even more violent collective events.

From the point of view of Freudian terminology man's personality was only partially formed, and consisted of ego and libido without the mitigating personality 'organs' of superego and consciousness. While the entire human personality existed as a potentiality at this time due to the well-developed hormonal and biochemical evolution of the humanoid, the overpowering needs of ego and libido, in this first dawning of man's sense of self, left little room for the emergence of the finer sensibilities characteristic of the formation of superego and rational consciousness in the developed human personality. In spite of the fact that from an overall perspective humanoid evolution is characterized by the distinct development of the 'female type,' physiologically and biochemically, in this first dawning of self-consciousness among humanoids, the evolution of a 'female type' in terms of psychic and cultural products would have been powerfully inhibited. The primal psychic power of ego and libido unmitigated by superego and rational consciousness would have lent unqualified evolutionary adaptive advantage to the 'male type' of the species, whose greater physical strength and more aggressive emotional makeup would be fulfilled only through force, violence and the overt acquisition of power over the lives and persons of others.

In other words, once the initial shock of consciousness of self had worn off, and the routines of fire control and containment and tool production had been established, humanoids in the lower Pleistocene settled in for a long period of psychic and cultural darkness. Manifestation of ego without full psychic development was expressed in libido aggression where 'male types' competed with each other for power with every conceivable means of opportunistic brutality and violence. Lower Pleistocene society was characterized by murder, and the physical abuse of all those weaker than the more aggressive and powerfully built males, including female mates and offspring. Naturally, this mode of behaviour accounted for the extreme limitations, at least by modern standards, in the proliferation of the humanoid species, a limitation that was even in contrast to the proliferation of the mammalian species who were far more stable in their collective and behavioural modes. The problem of consciousness of self and discovery of the ego in the upper Pleistocene and still not resolved to this day was that consciousness of self was an individual drive and experience and was in violent opposition to collective modes ingrained in the mammalian species as a whole and well established in the predecessors of Pleistocene man in his *Australopithecian* antecedents.

The 'accidental' discoveries of fire control and containment, the recognition of its use, and routinized tool production and burial patterns that led to the second efflorescence of humanoid evolution led to this kind of paradox. While the cognitive nature of these accidental discoveries demanded that man evolve even more complex social and cooperative patterns that allowed him to maintain his technologies, including rudimentary spoken language, the emotional consequences of these discoveries demanded that in every instance the stronger and hence most valuable members of the humanoid collective acted against the interests of these social and cooperative patterns. The result was a period of nearly 400,000 years of social anarchy where the humanoid species or genus as such constantly hovered on the brink of self-extinction, for with his increased knowledge and technological power lower Pleistocene man was not only a dangerous enemy to all members of this second efflorescence, but also the destroyer and enslaver of the remnants of the first efflorescence, *Australopithecian* man, whose lack of knowledge and technology made of him an easy prey for the unquenchable lust for violence he possessed. Resembling him in gross physiological form but not in knowledge of consciousness of self, collectives of *Australopithecus* who remained ignorant of the uses of fire and possessed no stone tools, became a scape goat for the violent lusts of Pleistocene humanoids who would otherwise have turned their lusts entirely against the members of their own collectives.

The pursuit and persecution of *Australopithecus*, while it continued to endanger humanoid evolution as such for a very long time, also had the evolutionary advantage of allowing the violence of the stronger male individuals of the collective to take forms that could be sanctioned by the females and other weaker members of the Pleistocene collectives. An institutionalization of the concept of the 'other' against whom violence was socially and collectively acceptable emerged, so that a margin of survivability for the humanoid species as such was obtained. If the need for violence had not been channeled against an identifiable 'other' outside of the humanoid collective, this violence would have been turned entirely inward upon the collective, allowing the total destruction of the females and the younger males who were the future of the humanoid species or genus. The intricate biochemical systemics which had evolved over 1.5 million years permitting the development and the potential for a 'female type' and a socially responsible 'male type' would have been extinguished, and evolution of life on this planet would have had to wait for another evolutionary experiment in some other species, mammalian or otherwise, in order to progress.

Thus, humanoid development emerged in the lower Pleistocene as a 'class system' where one group of humanoids preyed on, destroyed and eventually enslaved others who were weaker and slower in their attainment of self-consciousness and the control and containment of fire and tool making technologies. The ferocious results of unmitigated ego recognition would have been entirely suicidal to the race if the violence resulting from primitive personality formation could not have been projected on an 'other' outside of the collective, allowing the females and offspring of the collective to survive in relative safety. A socially responsible 'male type' thus eventually emerged during the period from the lower to the upper Pleistocene only in this context of a 'class system.' In this 'class system' which was in reality a pecking order, the stronger and more emotionally aggressive males were the most privileged, after them the female mates who belonged to them, then the offspring of these mates, and finally members of less developed humanoid collectives who were hunted by the males of the better developed collectives and eventually enslaved to serve this collective at its whim.

It is of course almost unthinkable to suggest that this kind of social system could even have achieved anything more than marginal survivability, certainly never any stable kind of social order, as all evidence from the sites of lower Pleistocene demonstrates to be the fact. Bit by bit however the development of this 'pecking order' among lower Pleistocene humanoids gave the collective the breathing space it needed to begin to fulfill the potentiality of the 'female type.' It allowed the females of the collective relative safety in which to express their natural feminine and motherly proclivities and to permit the longer and more secure time periods for the suckling and nurture of male offspring, especially, which led to the development of a less aggressive 'male type' in the collective. As the mother was allowed time and safety in which to care for offspring, the distance between the aggressive male father and the helpless male infant and child was mediated by the mother under more protected conditions. The infusion of male hormones in the male offspring was delayed to a period approaching puberty so that his cognitive abilities were able to develop more in tandem with his aggressive instincts. The pscyhological mediation between father and son by mother prevented, as well, early activation of defensive and hostile behavioural patterns between males from developing in the extreme way that had been characteristic of earlier lower Pleistocene society. Finally, in the upper Pleistocene, full personality formation of the humanoid type was achieved. The lengthy care of mother for son resulted in an internalization of the mother figure in the male and, of course, also female personality, where the power and strength of the mother figure held a critical psychic value in mediating between ego and libido, similar to the way in which the mother mediated in the social experience between father and son, or father and offspring in general.

As a matter of natural consequence, rational consciousness emerged then in the human personality as a means of balancing, arranging and calculating these highly complex intra-personal and inter-personal relationships. Man's psyche became the mirror of his social environment which arranged for his protection on the physical, ecological, psychological and social levels all at the same time. Man's consciousness was no longer bound merely by a sense of self or ego which could only find its fulfillment in acts of aggression and violence against all 'other,' and man's collective experience no longer resembled the unpredictable roller coaster of manic depressive whims in constant conflict that had been characteristic of lower Pleistocene society. Mankind had reached the threshold by which he was able to accomodate the needs of his own ego with the needs of others. He had reached the plateau where he was able to fulfill his own ego while submerging libido with the assistance of his superego, and to adjudicate this submergence with the aid of an emerging rational consciousness.

The emergence of the rudimentary hieroglyphics found on the walls of the caves of Lescaux near Dordogne in France in the upper Pleistocene was evidence of man's ability in his third efflorescence, to picture or model himself as a whole being in relation to the collective and to the universe, which was in turn a model of the collective. Man was no longer merely conscious of himself in isolation, he was conscious of himself in relation to the collective, to the intricate sets of relationships which were the culmination of all mammalian evolution. Thus, the statuettes of pregnant females found in upper Pleistocene sites did not represent the transition from a patriarchal society in the lower Pleistocene to a matriarchal society in the upper Pleistocene as is sometimes believed; rather these statuettes represented a charm or talisman of religious significance which was used as a symbol of the psychic, cultural and social development of upper Pleistocene man and which was held for religious and magical purposes to ward off regression to the more primitive and far more miserable state that had been known in the lower Pleistocene. Just as the mother was the mediator between father and son, and superego the mediator between ego and libido, so this talisman was the mediator between lower and upper Pleistocene, the religious protector against the descent into darkness on whose brink upper Pleistocene man hovered and which, even today, the more cautious of us, even in the 4th efflorescence of humanoid development, also fear to redescend.

In his famous work *Civilization and Its Discontents*, Sigmund Freud presented the thesis that man at the close of the Stone Age underwent a psychic transformation, based on two dynamic factors, both of which centered on the development of a superego in this phase of mankind's development. Development of the superego, Freud contended, made the individual male conscious for the first time of the fact that he had a father, and in the violent internecine struggle characteristic of Pleistocene man, the slaying of the father was wont to produce guilt. This guilt ensured eventually the relative security of a single leader whose inviolateness symbolized the recognition of the 'father' by each individual male. The other dynamical factor whose causes are somewhat more obscure in

Freud's work was jealousy of the male for the fertility and child-bearing ability of the female, which jealousy suppressed in the libido of the male formed the drive for males to create all the artifacts and devices that are characteristic of civilization, such as religion, art, law, science, architecture, music and etc. It was Sigmund Freud who made what is perhaps the greatest scientific discovery of modern times, for Freud analyzed the psyche of man in terms of its major organs, its ego, libido, superego and consciousness, albeit from an empirical rather than a historical or evolutionary point of view, from material he encountered as a practising physician, and it was in this role that he became the founder of modern psychology, psychiatry and psycho-analysis.

From the empirical evidence he encountered, Freud developed a psychological emphasis on the male rather than the female psyche, which from a historical point of view is entirely valid, as it has been primarily the psychic evolution of the male type throughout the latter efflorescences of the humanoid species that have accounted for the various characteristics of these efflorescences. Freud's major thesis was that in early childhood at about the age of five years the human male developed his ego in terms of his genital awareness; he became in Freud's terms, "penis proud" while the female's psyche at this age was evolved by an awareness of the absence of the male genitalia, developing a syndrome he characterized as "penis envy." From the historical point of view we may say that it was the consciousness of ego that, given the minimal conditions of lower Pleistocene society, comprised, in Freud's terms, "penis pride." This leads to, in the 'male type,' the evolutionary expression of ego through libido, or aggressive drives that have as their base a powerful male sexual component. The envy of the male genitalia in particular and the male in general by the female of the species is a minor characteristic of civilization, as evidenced by Freud's thesis that it is not envy on part of the female for the male which creates the products of civilization, but the envy of the male for the female. The fact is that female envy, as such, plays a minor role in cultural and social development in civilization because, by comparison with the quantity and quality of ego and libido possessed by the male of the species, the ego and libido of the female is relatively insignificant.

In other words, so powerful is the drive for ego-realization by the male of the species that once he is aware that the female is able to perform functions that he cannot, he strives to assimilate in whatever way he can these abilities that do not rightly belong to him. Since he cannot acquire these creative functions in any direct way, he displaces these primary functions into the secondary ones of a cultural and social nature. Thus it was not actual envy as such which motivated the male of the species to acquire the creative functions of a secondary nature which resulted in the products of civilization, but the constant factor of his overwhelming ego needs which drove him to find realization of himself through mastery of any particular function which he was able to imagine or conceptualize.

With the evolution of superego during the time of the upper Pleistocene, the impetus to find a sense of mastery through creative functions that were in fact not only harmless to the collective but also positively helpful to it, the production of cultural products as an expression of ego consciousness driven by libido instincts and aggression, would have had a highly positive sanction in the male psyche. Such a channeling of ego-recognition and libido drives would have had the full approval and encouragement of the females of the collective as it most certainly would have abetted their survival and security, and such approval and encouragement would have served to strengthen the development of both superego and rational consciousness among the males of the collective. This process produced the first 'social contract' in humanoid development, the 'social contract' characteristic of humanoid evolution in its third efflorescence.

Within the strictures of such a 'social contract,' the collective and behavioural modes of the males which had been stabilized during the long history of mammalian evolution, and destabilized during the 2nd efflorescence of human evolution in the lower Pleistocene, had an opportunity again of finding a certain degree of homeostasis or equilibrium. It was not necessarily simply the recognition by the male that he had a father and that aggression towards the male parent produced guilt in the male that enabled a relatively stable form of male leadership to develop in the upper Pleistocene. Within the strictures of the 'social contract' at this time, it was no doubt the recognition of the mother by the male rather than the father which allowed this relative stabilization of collective modes; for the relationship of the young male to his mother under the improved condition of the upper Pleistocene allowed full development of the superego without which no such stabilization would have been possible. The complex emotions of the male which drove him both to please his mother and to master her and her particular functions at the same time was the instrument by which such a 'social contract' was achieved. The acceptance of a male leader or leadership of the collective was a natural outcome of males who had undergone this kind of psychic development.

The feelings of aggression towards a male leader or leaders by other males would not have been stilled, but only mitigated by the mediation of the mother who became, as well, internalized in the male psyche as the superego. Due to the mediation of the mother in the social complex and in the internalized psyche, the overwhelming impulses of male ego and libido to dominate at all costs were muted, since these drives were to a large extent sublimated or channeled in directions that were in fact more fulfilling than simple drives to dominate which were after all, in most respects drives that would have been impossible to realize on the part of all but a single or, at most, only a few of the males of the collective.

With the evolution of the humanoid species during the 2nd to the 3rd efflorescence, where the male offspring was allowed a longer, closer and more secure relationship with his mother, the single need to dominate other members of the collective became complicated by another need, the need to please the mother. The greater period of time in which the male system was infused with female hormones, due to this period of childhood grace, made him more prone to extend this need to please his mother to other members of the collective. With the infusion of female hormones into his system at an early age and throughout a genuine period of childhood grace in later years the male would be able to model motherly behaviour towards the younger and weaker males of the collective. This same process would also condition the male in later life to continue to model himself in a childlike role towards a male leader or leaders of the collective, without the single minded impulse to enter into combat with him for the purposes of destroying him and taking his place as the single dominant male of the collective. The need to dominate took a safer and more willing object of desire—the role of the female. In this way the impulse to domination or power became sublimated and the products of civilization could be created.

The value of such a 'social contract' in the first instances was appreciated by the male in a very pristine way. Recognition of the internalization of the female as the superego by the male seemed to have almost a supernatural aura. This internalization allowed a device by which the collective could be protected and could be proliferated without the extreme anxiety that such stability portended during the lower Pleistocene. Acquisition by the male of a superego meant that he no longer suffered anxiety and conflict in not perpetrating violence on the collective. It allowed him to attain a serenity which left room for his own ego expression as well as the survival, if not the ego expression, of others in the collective. Considering the darkness from which he had so recently emerged, such a happy concatenation of circumstances seemed nothing less than miraculous to him. In addition, such a development was entirely due, not to any material development, but to a psychic development, an intangible addition to his person which he necessarily recognized as such. He had attained a consciousness of his internal self, and this consciousness was necessarily perceived as a 'picture' of himself. Religion, art, and written symbols emerged simultaneously from the psyche of man in the upper Pleistocene.

Consciousness that there was something 'inside' man that governed him, providing him with stability, security and order, was tantamount to recognizing that there was something 'outside' man that was responsible for these fortuitous benefits. Since the nature of internal self was intangible, the location of that self also became intangible, and its presence, in this first dawning of man's moral consciousness, was recognized as a spiritual power that infused all things, not just the person of the individual. A sense of deity was attained that permeated all things, animate and inanimate, man and animal, self and the 'other.'

In all of this, in applying Freud's psychological theories to the evolutionary development of man, particularly his emphasis on the development of the male psyche as opposed to the female, the purpose is not meant to devalue the importance of female psychic development in favour of the male. On the contrary, Freud's scientific analysis of personality is very much in keeping with the actual facts and circumstances of mankind's evolutionary development in which psychic development plays such a critical role. The 'female type' of the humanoid species is stabilized early in humanoid development and the stability of this type is a determining aspect of mammalian evolution in general. Given the two 'x' chromosomes associated as genetic features of the humanoid female and the more central association of this type with the late special development of the upper nervous system or cerebellum, the 'female type' of the humanoid species is from the beginning of its development freed from the potential problems of ego and libido that the 'male type' is beset with. The 'male type' of the humanoid species, containing within his genetic inheritance both an 'x' and 'y' chromosome and his association with the lower nevous system or medulla as the vestiges of the pre-mammalian mandate for individual as opposed to group survival, contracts severe problems with ego and libido when these psychic developments occur in humanoid evolution. This is not to say that females do not have egos and libidos and may also contract severe problems in adjusting to this collective psychic evolutionary phase.

Many of Freud's most interesting cases are with the ego and libido problems of females rather than males, if only perhaps because these problems, being less severe in females than males, were more amenable to his pioneering efforts to cure these problems. If in the early stages of childhood, the ego of the young female is damaged by a violent, intemperate or indifferent relationship with the male parent, severe problems of 'penis envy' may occur which will continue to distort proper personality development throughout the lifetime of that female. Or, if in the early stages of childhood the young female has an inadequate or deranged relationship with the female parent, she may develop a severe case of the 'electra complex,' which in simple terms means a highly aggressive and hostile attitude towards the female parent which may, in later life, severely affect the normal development of both personality and the ability to form stable social relationships in the life of that female.

The complementary complexes or inadequate personality formations that occur where the young male is concerned are damaging of the sense of 'penis pride' by a taunting, indifferent, or violent relationship with the male parent. Or, an inadequate relationship with a female parent when the male is young will result in a severe 'oedipus complex' on the part of the male in later life, which means, in simple terms, a hostile and aggressive attitude usually both towards the male and female parent and all persons who in later life become extensions of these roles. However, the social effects of deranged parenting towards the male have far more serious consequences than in the case of the female as can be seen from the fact that the vast majority of social crimes and general violence in every known society are committed by males rather than females, while there is no reason to suppose that there has been greater deficiencies in the parenting of males than females from a random, statistical point of view. However inadequate parenting occurs in both cases of males and females, the social consequences of inadequate parenting for males has far more dire social consequences than inadequate parenting for females. Due to the general evolutionary trend in both mammalian and humanoid species the 'male type' of the humanoid species is more complex and more unstable than the 'female type,' and the vicissitudes of humanoid evolution after the 2nd efflorescence of humanoid development are a measure of the circumstances in which the psyche and personality of the 'male type' rather than the female are able to develop.

In the evolution of the humanoid species after the second efflorescence of humanoid development in the lower Pleistecene the 'female type' remains a relative constant while the 'male type' is the relative variable. Whatever the psychic and cultural accoutrements the 'male type' was able to acquire since the advent of the lower Pleistecene, he could never escape from the special burden which fell on his shoulders as a result of general species and mammalian evolution—the need to protect not merely himself but also the female of the species and her young. This special need results in a plethora of secondary drives and impulses which, when unchanneled or improperly channeled, result in social disorder and potential catastrophe for the humanoid collective, and when properly and constructively channeled, as begun to happen in the upper Pleistecene, result in a complicated social order and civilization inconceivable to earlier efflorescences of humanoid development and hitherto unknown by life forms on this planet. The 'male type' can no more escape from his special set of drives and impulses than the 'female type' can escape from her ability to conceive, bear, and suckle offspring, for the special needs and drives of the 'male type' are as much an aspect of his gross physiology and specialized biochemistry as childbearing is for the 'female type.' Again, this is not to suggest that the 'female type' cannot partake in the production of the psychic and cultural objects of civilization. Certainly this is not what was meant by Freud, and is not what is implied in the evolutionary analysis of humanoid development. Nonetheless, the fact is, at least in the first instance of evolving civilization among humanoids, much of the impetus for the production of the psychic and cultural objects of civilization were the result of the sublimation of aggressive instincts which are predominantly more characteristic of the 'male type' than the 'female type.'

As civilization evolved during the 3rd efflorescence of the humanoid species and continued to evolve in the 4th efflorescence of which we are today the pioneers, collective modes among both 'male' and 'female types' became stabilized, and the distinctions between 'male' and 'female type,' after certain points of biological maturation, became more and more blurred. From this blurring of distinctions in higher forms of evolved civilization the 'human type' in contrast to the either the 'male' or 'female type' *per se* emerges, where on the basis of rational consciousness alone, both psychic and cultural products of civilization and more refined modes of collective behaviour may be instigated as a matter of individual choice rather than sexual necessity. Such a development however must be considered a primary characteristic of the late stages of the 3rd efflorescence or the 4th of humanoid development which evolved during the period from 2000 years ago to 20,000 years ago, the period we designate today as the period since the onset of 'post-historical' or 'historical' time. It is most unlikely that individual choice as opposed to sexual necessity determined the genesis of the products of civilization in the upper Pleistecene which were primarily the result of the evolution of the 'male type' in this period.

On the other hand, it is also unlikely that modes of collective behaviour during this period were inspired by drives peculiar to the 'male type,' but were the result of the inspiration of the 'female type,' whose mode of behaviour was copied by the male and allowed to flourish as a general mode in the collective once the 'male type' had internalized its mechanisms in his own psyche in the form of superego. Thus, it is interesting that statuettes of pregnant females found in the sites of the upper Pleistocene have their faces covered with homogeneous lines or etchings, resembling perhaps hair, which symbolized the fact that although the 'female type' was deified it was deified without individuality. Thus it was the 'female type' as such who was deified by the humanoids of the upper Pleistocene, rather than any particular female, for the valuation of the 'female type' was the instrument by which superego was internalized, and the mediating role of woman institutionalized, the two critical changes in the development of upper Pleistocene humanoids which allowed the further development of the humanoid species in its third efflorescence.

The most significant controversy in the history of psychological thought centers around the theories of Sigmund Freud and those of his leading and most famous disciple, Carl Jung. While credit must be given to Jung as the first thinker who attempted to put psychological theory in the context of man's full evolutionary development, his theories have often carried little credence except with those who propound quasi-scientific theories of race development and differentiation as they believed, on totally inadequate examination of the facts, it to have developed in post-historial time. If we ignore the racist implications of Jung's work, which in fact were never propounded by him as any overt sort of ideology but only taken up by others who used parts of his theories for their own purposes, and instead see the general trend of his work in terms of the kind of evolutionary development of mankind which we have been discussing here, many aspects of humanoid evolution become clearer.

Jung maintained that mankind possessed not only an individual unconscious but also a 'racial' unconscious. As he expounded the nature of this racial unconscious, it was seen to be comprised of myths and even fairy tales which are the common lexicon of peoples in historically developed civilizations of Europe, the Middle East and Asia. Jung illustrated that there were a seemingly finite number of actual myths common to the peoples of these civilizations, where the form differed slightly from region to region and from time to time while the central drift of a given myth remained constant. It was part of Jung's central thesis that the mind of man had a proclivity to think in terms of these finite number of myths, just as the mind of man had a proclivity to organize his experience around the central psychic organs of ego, libido, superego and consciousness, whose dynamics lay in the childhood dramas of parent and child, and male and female sexuality, common in an absolute way to all mankind. Jung maintained with Freud that humans were born with the psychic organs of ego, libido, superego and consciousness much as they are born with arms and legs, and that the prognosis for the healthy development of these psychic organs was as dependent on the proper environmental psychic nutrition much as organs of gross physiology were dependent on food nutrition for their proper development.

Taking as he did with Freud such a deterministic view of man's psychic constitution, Jung ventured to speculate that man was similarly 'born' with a reservoir of race memory or myth which comprised the history of human evolution, inborn and innate, in each human psyche. The philosophical wrangle that Jung found himself embroiled in for the duration of his career and which led ultimately to a serious break with his mentor and teacher, Freud, was that while an innate individual psychic constitution was of a structural nature, an innate racial psychic constitution was of a processual nature. In other words, while individual psychic makeup could be understood in terms of basic components or psychic-like organs, the nature of the racial psychic makeup remained far more vague and indeterminate, although Jung ventured to designate this racial psychic makeup in terms of what he called 'archetypes.' 'Archetypes' were the seemingly finite number of mythic structures which were present in the individual as a result of the individual's racial memory and which for any given individual had an infinite number of possible variations on the same basic, finite themes, due supposedly and apparently to the actual 'race' of which that individual was a proper member.

The conflict between Freud and Jung in psychology was similar to the conflict between their German contemporaries and colleagues in the fields of physics and mathematics, Einstein and

Heisenberg. Like Freud, Einstein was a humanistic rationalist who believed that the laws of nature could be submitted in all cases to rational analysis and conversely that that which was not capable of rational analysis was not in the sphere of science proper. Like Jung, Heisenberg was a mystic rationalist who believed that a rational model could be used, if not to explain, to express, the mystical and the irrational. The legacy of Heisenberg has left us with a rational model of a physical universe that is uncertain and indeterminate, an orderly conception of that which is essentially disorderly and chaotic. Jung's legacy has been similar, for he maintained that we could remember, in fixed patterns, that which we could never have possibly known.

However, Jung's theories become more explicable if we examine the nature of the myths he described in the context of the evolutionary development of mankind. The myths that Jung deals with are largely myths of the libido, as he himself explained, and from the individual material of protoschizophrenic patients which he used, these myths reveal a conflict between ego-realization and the overwhelming onslaught of libido aggression. The familiar figures of these myths are animals, natural elements, or natural occurrences and catastrophes, such as earthquakes, rains, landslides and so forth, and the events in which these figures occur symbolize such psychological structures or archetypes as rebirth, abnegation of world or self, sacrifice, homomorphism and incest. From the point of view of humanoid evolution in its 4th efflorescence, personality degeneration of the kind from which Jung's material is derived, suggests nothing more or less than the fear of reversion of the developed humanoid personality of this time back to the conditions of humanoid personality during the 3rd, 2nd and even 1st efflorescences of humanoid evolution.

Humanoid evolution in its present stage exists where rational consciousness is based on a prior development of superego consciousness and where that in turn is based on a prior development of ego-consciousness expressed through libido aggression. The development of the human personality in its present state is a result, not of inborn structures *per se,* but on a developmental process through which these structures are realized and crystallized under appropriate circumstances of parenting and socialization. If these circumstances are lacking for any particular individual in its developmental stages, the final development of that individual to the present culturally idealized stage of human personality development is by no means a necessary development. Any human individual whose personality development is inhibited due to deranged parenting or inadequate socialization runs the risk of never attaining the present culturally idealized state of human personality development, a development characteristic of the late stages of the 3rd efflorescence, or the 4th, in humanoid evolution. Depending on the severity of the derangement of parenting or other circumstances of socialization, the human individual may only achieve the stage characteristic of the 3rd, 2nd or even 1st stages of humanoid efflorescence.

While from the point of view of the late stages of the 3rd efflorescence or the 4th in humanoid development, these prior stages may be designated as 'schizophrenic' or 'psychotic,' or 'manic depressive' or simply 'insane' psychological states, these are socially or phasically relativistic judgements when seen from the overall point of view of humanoid evolution. The inability of so-called 'schizophrenic,' 'psychotic,' or 'insane' individuals to adapt to the present collective modes of behaviour, to be socially and individually disfunctional, is understandable from the point of view that humanoids possess both instinctive and developed modes of collective behaviour. If an individual has only attained the personality development characteristic of say the 2nd efflorescence of humanoid development and is necessitated to behave in a collective mode with humanoids existing in the 4th efflorescence of humanoid development, that individual will appear to be, and in fact will be, socially disfunctional.

The point here of course which Jung neglected, no doubt due to the inadequate knowledge of humanoid evolution that existed in his time, is that the parenting and patterns of socialization that go into producing a personality developed to the stage concomitant with the 4th efflorescence of humanoid evolution is not an absolute, but a matter of evolutionary process. 'Mental health' or 'mental sickness' is not an absolute criteria, but an evolutionary one. This becomes clear once we recognize that man's evolution is of a primarily psychic nature, not of a gross physiological one. Thus the kinds of phantasies or myths characteristic in individuals who are schizophrenic or 'insane' by modern standards and which in the cases which Jung primarily deals, are libido myths; these phantasies and myths would be entirely 'normal' among individuals in the lower Pleistocene, where patterns of parenting and socialization were ill-formed and undeveloped by modern standards, where neither superego nor rational consciousness had become institutionalized or internalized as modes of collective behaviour.

The fact that myths or phantasies characteristic of the lower and upper Pleistocene are 'remembered' by modern day individuals results from the fact that, due to the nature of humanoid evolution, there are priorities in the development within each individual of the various psychic organs. As the young humanoid develops from infancy to childhood, ego is the first psychic organ that develops, as Freud maintained. Libido is the means by which ego-realization is achieved. If inadequate or deranged parenting occurs at this stage, and the inadequate and deranged parenting is reinforced by inadequate secondary patterns of socialization, it is entirely possible that the mother will not be acknowledged as a mediator between offspring and father, and superego will never be properly internalized. Even if superego is internalized, inadequate or deranged parenting combined with inadequate collateral modes of socialization may prevent the ascendance of the superego within the individual psyche. The individual may be left forever behind in a stage of development characteristic of the 2nd or 3rd efflorescence of humanoid development. That the figures of myths characteristic of ancient folklore which are largely libido or superego oriented repeat themselves in the phantasies of individuals with inadequate personality formation in modern times is also not a matter of mystery, as the external environment in which we live, with animals, natural elements and natural catastrophes is hardly very dissimilar from the natural environment of 500,000 years ago.

To an undeveloped personality who has been unable to conceptualize in terms of the symbols of superego and rational consciousness these figures naturally express the level of individual personality development. In addition, we are talking about individuals, not in the lower or upper Pleistocene, but in modern times, who in spite of inadequate personality formation have learned the strictly cognitive forms of verbal communication and no doubt have learned to read written language in a common educational context; albeit the emotional affect of the material learned cognitively is not integrated properly in the total personality of the maladjusted individuals, the common lexicon of culture where these myths would be learned would be at the disposal of those individuals for their particular maladaptive personality proclivities.

Nonetheless, and in spite of this attempt to explain Jung's unique theories in a rational way, there are still interesting implications that remain and that are quite inaccessible to entirely rational explanation. The fact remains, and is indeed not quite explicable, that during four efflorescences of humanoid development over a period of several million years there are a seemingly finite number of myths of a relatively common form that the human mind continues to reproduce. While it is true that these myths have various senses of significances depending on the period of human development during this period, this variation of significance is almost dwarfed by the overwhelming import of the fact that these myths nonetheless continue to be reproduced. The inevitable implication is that not only is the structure of the human mind of a largely predetermined nature, given the first development of the humanoid species, but that human thought itself is predetermined.

To put it another way, we can see already at the advent of *Australopithecus* several million years ago that although the structure of the human mind as it exists today in its properly developed form was not an actuality, it was a potentiality. That potentiality among the *Australopithecus* collective was as real as the potentiality of any infant humanoid born in society today. The difference between the *Australopithecus* individual and the infant today is the time of evolution during which collective and psychological modes were developed to allow that potentiality to become almost an automatic actuality. If the same thing is true of human thought then we must regard thought itself as tangible and 'organic' as the biological and biochemical structures which produce it.

The conflict between Heisenberg and Einstein in the fields of physics and mathematics and between Jung and Freud in the field of psychology in 20th century Germany was presaged by the philosophical conflict between Hegel and Marx in the 19th century in the field of social philosophy. These conflicts represent the parameters of the crises and challenges which face mankind at the present time in the 4th efflorescence of his development, and the outcome of these conflicts may well determine not only the ultimate success or failure of the 4th efflorescence of humanoid development, but also the very existence of the humanoid species or genus on this planet in any form. For, as in the onset of the 2nd humanoid efflorescence, mankind stands, posed precariously on the brink between the hazards of extinction and the opportunities of proliferation.

In this period, all the complex collective modes and technologies of mankind's vast resources are organized on the various sides and edges of these conflicts spawned in the cultural developments of 19th and 20th century Germany, posed for conflagration and self-extinction or further progress and development of the humanoid species. Out of a melange of classical Platonism, didactic Aristotelianism and the new scientific and philosophical speculations and discoveries in 19th century Germany, Hegel proposed that the evolution of mankind, his history, thought, and material development, was an evolution of dialectical thought or pure energy. The synthesis of such a dialectic, Hegel held, was in the evolution of a pure collective or racial group which would be entirely self-conscious of its ideal development. With a more pragmatic sensibility, but with much premature scientific theorizing, Marx stood Hegel on his head and maintained that the evolution of mankind was entirely material and that thought itself was only an ephemeral reflection of the material conditions of mankind's evolution at any particular point in time. Marx maintained that mankind's progressive evolution could be measured in terms of which members, and which percentage of the members, of the human collective controlled the economic means of production of the collective.

The final synthesis of mankind's evolution in Marx's view was the diffuse control of the means of economic production by all, rather than some, members of the collective, this diffuse control being guided and informed by those individuals most clearly aware of mankind's evolutionary path and destiny, those individuals who would band themselves together as members of the Communist party. From the perspective of hindsight, informed by the work of Freud and Jung, and a more exact anthropological grasp of the actual datum of Marx's evolutionary history, we can see that it is not the status of control of economic means of production that actually measures the progress in man's evolutionary history, but the status of the drive for control *per se*. It is the evolutionary development, and then the harnessing, of man's ego and libido which are the central constants which measure mankind's evolutionary progress, while such componants as economic means of production or technological tools of production are merely variables. Significant as economic means of production or technological advances in production of goods may be in some circumstances, they are not real measuring sticks of man's evolutionary progress. Diffuse control of economic means of production among a collective where psychic evolution of the collective is at a regressive stage of development is an occurrence far more common and more likely to occur than the idealized state imagined by Marx, who neglected, partly, of course, due to the prematurity of his scientific theories, the underlying psychic dynamics of human development. On the other hand, Hegel's theories which held that cosmic thought was the essence of human evolution, although perhaps more philosophically pure than Marx's theories, remains too far from the dynamic realities of human biological and biochemical development, with its concomitant material accoutrements characteristic of varying stages of this development, to be of any real utility as a theory of human evolution.

The fact of the matter is that human evolution is both a material and an ideal development. The foundation of any theory of human evolution must rest firmly on the foundations built by the great English scientist of the 19th century, Charles Darwin. Out of the matrix of species development on this planet, mammalian evolution was a unique biological development, of which humanoid development is a further and more complex differentiation with a phenomenological adaptation that is at the same time both material and ideal or spiritual. Once humanoid evolution had occurred out of its mammalian base, the upright-standing two-footed humanoid possessed psychic potentialities which took several million years to develop to its present state. For all intents and purposes, the development of the gross physiology of the humanoid species ended at the very point where it first began, and from that point onward evolutionary progress in humanoid species was of an ideal or psychic nature. The mediator between the evolution of man's gross physiological development and his psychic development lay in the complex biochemical and neural biochemistry which lies, as a matter of substance, somewhere between man's material and his ideal or psychic form. In the first instance, this biochemistry is produced by gross physiological features and organs, but it is reactive to thought and psychic development, assists in its realization and is no doubt acted upon and changed by thought and psychic development, just as it acts upon and changes thought and psychic development.

Thus, in seeking cures to what we term mental illness where individuals have regressed to periods of humanoid evolution of the upper or lower Pleistocene, both Freud and Jung found that these illnesses were not usually amenable to therapies that were entirely of a purely psychic nature. Latter day psychiatrists have found that biochemical treatments by drugs that affect the hormonal and neural biochemical systems are far more effective in combination with psychic therapy than psychic treatment alone. In a majority of cases, in fact, biochemical therapy alone of a maladaptive individual allowed to rejoin the normal collective has positive therapeutical results. In the context of Hegel's general argument, psychic development is of an 'organic' nature, but this 'organic' nature of thought is not an abstraction, it is in fact exactly analogous to an organ of a gross physiological nature. It has an entire ontogenesis, a conception, a period of lability and growth, reaching a rigidity in structure in maturity, and of course a death with the physical demise of the individual who possesses it. The disfunction of psychic growth is thus exactly analogous to the disfunction of growth in any organ of gross physiology.

If in its early stages there is severe inhibition, in the later or mature stages of that organ that organ will be deformed and, short of performing actual regeneration of that organ, which is by present medical methods impossible, that organ will remain malformed for the life of the individual. Freud clearly had this model of psychic development in mind, in terms of the development of 'psychic organs,' and his method of psychoanalysis in which the ill patient was allowed to relive and hence regenerate these 'psychic organs' was based on this model. However where psychic development is seriously impeded there are no doubt fistulas and contusions in the biochemical 'organs' that mediate between organs of gross physiology and psychic organs or pure thought, and these fistuals and contusions are increasingly amenable to biochemical therapy as drugs are developed that more perfectly resemble the structure of these biochemical 'organs' or systemics.

Nonetheless, in all of this it is the question raised by Jung that is the most critical one of all for an understanding of man's evolutionary development. Synthesizing the works of Hegel, Marx, Freud, and even Heisenberg and Einstein, and founded firmly on the work of Darwin, the central issue that is so critical and so novel in Jung's work is that of man's common mental evolution, the racial unconscious or the universal existence of archetypes in the minds of all men. If the evolution of man's thought is as fixed and predetermined as any of his gross physiological organs, in what way then is his thought free? Does the evolution of human thought have any more general relevance to the universe then than the accidental and particular evolution of, for example, his upright, two-footed stance? Is man's ability to think a particular aspect of evolutionary history then, like his upright stance, or does it have a more general relevance?

Is this concept of the general relevance of man's thought, his supposed ability to analyze the workings of the universe and himself nothing more than the latter day expression of the needs of his overwhelming ego? It is of course only in the posing of such questions that we may even hope to answer them. In the act of posing such questions man attains the degree of development characteristic of his later efflorescences, the consciousness of his own rational powers, independent of psychic organs which are prior to rational consciousness in humanoid development, such as ego, libido, and superego. With the attainment of rational consciousness, man is able to transcend the particularity of his 'accidental' evolutionary development and to grasp the generality of evolution as such.

Chapter III

THE COLLECTIVE MIND EMERGES TO DISPLACE SUPEREGO IN MODERN SOCIETIES

The fact of the matter is that the psychic evolution of man since the lower Pleistocene, in the 2nd efflorescence of mankind, to the rational consciousness of his psychic evolution that is characteristic of man's 4th efflorescence in post-historical time has by no means been either smooth or continuous. Even in the flowering of Greek culture in 500 B.C. where distinct valuation of rational consciousness became institutionalized in human collectives, the Greek society in which this development occurred was a state built on the slavery characteristic of the lower Pleistocene. Modern American society which held the proliferation of rational consciousness as the *raison d'etre* of an international collective mode was also in its first instances a slave state of distinct proportions, where collective libido was kept in check by the awesome enslavement of one human group physically and culturally distinguishable from the group or race it enslaved, a mode of behaviour even more similar to lower Pleistocene society than ancient Greek culture where slavery was never organized on the basis of distinct racial characteristics.

It is also a matter of common understanding that improved methods in communication and transportation which have been in general characteristic of post-historical society and have in particular rapidly escalated throughout the world only during the last 200 years have not resulted in standardizing collective modes central to the 4th efflorescence of human development. On the contrary, they have as often been responsible for inhibiting humanoid development in this efflorescence as promoting it, and as such these improved methods of planetary communication and transportation remain one of the critical reasons why mankind as a whole continues to hover on the brink of universal regression and even universal self-extinction. In evolutionary development 'more' is not necessarily an indicator of 'better.' In fact, such is the nature of evolutionary development and efflorescence of an advantageously adaptive type that it is the 'accidental' change in a single individual in a single small collective that ultimately results in advantageous adaption leading to the efflorescence of the type. Just as when efflorescence is achieved through force, as with the efflorescence of the single-minded aggressive dinosaurs, so evolution that is 'forced' may lead in the end to the entire extinction of a species or even a genus.

Improvements in mass communication and transportation that serve to proliferate during a short term a maladaptive form of a species, or maladaptive collective modes of behaviour, may result in the widespread destruction of an entire species or genus with all of its individual potential for adaptation as well as maladaption. In addition, the exponentially increasing proliferation of mentally ill individuals and sociopaths in all the industrially developed societies of the world and, in particular, those of both North and South America are indications of the developmental abyss on which mankind today continues to be perched. In these free-wheeling capitalistic societies, the evolutionary race is on. The question is whether free-wheeling economic strategies practiced on a mass basis will provide ultimately the material resources by which maladaptive and regressive individuals may be helped on an effective, mass scale before the number of these maladaptive and regressive individuals sway the entire social balance in a degenerative direction, subsuming all the advances of the 4th efflorescence of mankind under the aegis of some more primitive collective mode of behaviour.

In Communist countries of Europe the question is whether individual self-realization will lead to more adaptive forms of collective behaviour or whether they will degenerate to collective modes of behaviour where rational consciousness is abandoned in favour of superego mediation characteristic of the upper Pleistocene. In such a circumstance, leadership of a mindless and unquestioning collective will be up for grabs by individuals of greater or lesser degrees of degenerative personality types, such as occurred in the Soviet Union during the time of Stalin. In the Third World of Africa and Asia, the question is whether these regions will be subsumed under the maladies of both the 'free' and Communist developed regions of the world, or whether they will be able to evolve their particular collective modes that will characterize them in the 4th efflorescence of humanoid development in tandem with the particular material and technological accoutrements they have borrowed from the industrialized countries of the world. There is no question but that mankind today hovers on the brink between destruction and progress in a number of variable ways and directions, but the further question is whether or not he has been there before.

While individual human life is very short, the life of the collective is very long, and what may appear as in sharp focus from the point of view of an individual in a single life span, from the point of view of the collective evolution the sharpness of this focus may seem dim indeed. From the point of view of humanoid evolution, what appears as catastrophe or crisis in a single life span may be no more than the infinitesimal points or blurred edges of a much wider and more continuous catastrophe or crisis.

Thus, the consciousness of self and the realization of ego through expression of libido that occurred in humanoid development in the lower Pleistocene, from the point of view of not merely mammalian development but from the point of view of development of all life forms on this planet, might have been a crisis or change as fundamental as the evolution of the lower nervous system which permitted the differentiation of vertebrate from invertebrate life hundreds of millions of years ago. The crisis which this first fundamental change portended for life on this planet where the aggressive instincts resulted from the development of the lower nervous system was mirrored in this second fundamental change in evolutionary development.

The aggression which resulted from the development of the lower nervous system, allowing the proliferation and efflorescence of vertebrate life forms, placed life forms in a continual crisis, for the aggression which permitted this proliferation and efflorescence at one and the same time placed these life forms in continual danger of causing the very extinction of these life forms. This contradiction in the development of humanoid species who attained a sense of self that could only be expressed through the aggression and violence of libido was reflected in the sense of self, or life of self, that was entirely limited by a sense or recognition of the death of the self.

All of the developments in humanoid evolution since this fundamental evolutionary change in the lower Pleistocene were adaptive changes which attempted to ameliorate or stabilize human collectives faced with this critical dilemma.

Given the extreme nature of this crisis, such adaptive and ameliorating changes leading to relative stabilization of humanoid evolution in the upper Pleistocene and in historical times have by no means followed a continuous or easy path, so much so that the adequacy of these adaptive changes remain even today very much in doubt. In pre-vertebrate species and in the bacteria and plant kingdom, the opposition between life and death is so blurred that it is often seemingly irrelevant, for it is not the individual of the species or genus that has life or death but the species or the genus itself. In the plant kingdom, there is regeneration of the individual plant and dormancy rather than death as such during intemperate or hostile seasons or conditions, while in the insect phyla 'communication' or 'thinking' as such results from biochemical compounds whose production is randomly produced among all individuals of the species and across time or generation in a way that makes any notion of life or life-consciousness inadequate. Among bacteria where reproduction is asexual, there is no clear differentiation between parent and off-spring, no way to distinguish in terms of gross physiology or even biochemical makeup one individual from another in a given bacterial species.

With the evolution of the lower nervous system in the vertebrates, life and death became matters for the individuals of the species rather than the species as a whole to contend with, and with the development of ego-realization in the 2nd efflorescence of the humanoid species, this crisis of general species development became crystallized, not merely as an aspect of material or biological development, but also as an aspect of the evolution of thought as a new addition in the evolutionary pathway.

Thus, during the 400,000 years from the lower Pleistocene to the upper Pleistocene and beyond to historical time, the problems for humanoid development continue to center around this all-important crisis. The development and recognition of the superego in the upper Pleistocene was an adaptive amelioration of this crisis, not a final solution, and indeed the same can be said for the development of rational consciousness in the 4th efflorescence of humanoid development. Whenever humanoid development progressed during the 400,000 years from the lower to the upper Pleistocene, where superego was able to develop and organize around itself these adaptive forms of collective and behavioural modes, these collective and behavioural modes remained always in opposition to the central nemesis of ego recognition and its expression through libido aggression in each generation and with the birth of each new individual of the collective. With the birth and growth of each new individual, the acute and painful self-realization of ego-consciousness would be repeated once again, with all the costs to the adaptive modes of the collective while this ego was expressed through hostile and aggressive libido.

In the earlier stages of upper Pleistocene development, or the late stages of lower Pleistocene development, simple recognition, at some point, of the superego by every individual of a given collective would have hardly provided the adaptive amelioration of the ego/libido crisis necessary to improvement in conditions for the collective. For superego to produce definite advantageous adaptation for the collective, individuals of the collective would have had to do more than simply acknowledge the existence of superego, they would have had to be able to actually organize emotion and behavioural modes around this recognition. For this to have occurred, a number of devices associated with superego development would have had to be learned by individuals of the collective. These devices would include fertility rites where the female was worshipped by the collective, acknowledgment of special individuals as priests, doctors and artists who embodied in their persons and special abilities the power of the superego. Having this 'power' of superego these individuals would come to represent a sublimated form of ego and libido expression, where the desire to know and to covet for oneself would be channelled into more socially advantageous modes of behaviour.

Another device would be the recognition of female kinship, as such, as opposed to male kinship, for the mediating power of the mother and the superego could be embodied in these series of relationships where ego and libido expression would again be sublimated into less aggressive and more socially creative forms of behaviour. However since adaptations that evolved from recognition of the superego were secondary rather than primary aspects of the central humanoid crisis, adaptions related to superego development in the upper Pleistocene, as opposed to those related to ego-recognition and its expression through libido aggression in the lower Pleistocene, were of a highly variable nature. Thus, while upper Pleistocene efflorescence was characterized in general by religion, art, medicine, written hieroglyphics and relatively well-formed spoken language, female kinship systems and the valuation of the mother figure, in particular, from region to region and from time to time, there were vast differences in the practices of these adaptive devices among different collectives.

Written culture and language as it evolved during the upper Pleistocene was a hedge against reversion to the behavioural modes of the lower Pleistocene. When culture was written and drawn it could be naturally more easily disseminated and taught to the young of the collective, so that each birth was not such a long-term gamble on the ability of the collective to survive the identity crisis of the off-spring once it gained ego recognition and expressed this recognition through violence and hostility in maturity. Early training of the young in all the learned devices which bound superego more or less firmly in the psyche of each individual member of the collective was the method by which the group learned to insure itself against its members. In tandem with the relatively safe childrearing periods granted humanoids in the upper Pleistocene, the artifacts and symbols of written and artificially produced culture inculcated into the minds of the young while they were still in this dormant ego and libido period was the central social engineering feat that ensured the survival of the humanoid species into a 3rd efflorescence.

During the flowering of the 3rd efflorescence of humanoid development the power of the human ego and its expression in libido aggression lay dormant. Human collectives everywhere during this period were preoccupied with the learning of the multitudinous devices that would eventually make the institutionalization of the superego and the mother figure a custom to be taken for granted among human collectives, rather than a task which absorbed the ego and libido of individuals in the acquisition of it, as in its early phases. At the point when the institutionalization of superego and the mother figure became a custom late in the upper Pleistocene, rather as the use of fire became eventually a custom rather than a new and difficult task to be learned in the middle Pleistocene, the humanoid species was ready for its next evolutionary leap into its 4th efflorescence. When the institutionalization of superego and the mother figure had become a custom, no longer absorbing the affect of ego-recognition and libido aggression, in such pursuits as medicine, art, priesthood, acquisition of rudimentary written language and the formation of matrilineal kinship bonds, ego-recognition was once again freed and gained ascendancy, although now, in post-historical time, in a different form. Ego-recognition and libido aggression became identified with the collective, and collective, or historical, consciousness was born. The male leader of the collective or his sometime surrogate, in the form of the wife of a dead leader or the daughter of a leader who had no sons, became the focal point of ego-recognition and libido aggression. Whereas in the lower Pleistocene ego-recognition and libido aggression had produced the recognition or 'naming' of the individual, in the 4th efflorescence of humanoid development ego-recognition and libido aggression produced the recognition or 'naming' of the group as such, and nations were born.

Ethnological data collected during the 20th century has produced a plethora of information about human groups in isolated parts of the globe, such as Polynesia, New Guinea, parts of Africa, southern Asia and among native Indian groups of North and South America who have or had remained firmly in the 3rd efflorescence of humanoid development. This data and these groups have remained the object of study of the field of modern anthropology, although with the ever increasing incursion of modern industrial societies on these groups in the forms of political and economic colonization, their number today has dwindled to a mere handful of collectives with increasingly fewer members. While they remained relatively isolated in the first part of this century, studies have shown that these groups remained for the most part tribal, leaderless and without national or collective consciousness. All of these groups possessed doctors of a sort, priests, artists, often rudimentary written languages, and highly complicated kinship systems whose rules provided the basis of marital, property and other jural decision making. These groups found in this century had all been geographically isolated by sea, mountain or jungle, desert or icy steppe from those collectives who had made the leap into post-historical time, humanoid development in the 4th efflorescence.

As can be seen clearly from their increasingly dwindling numbers, once exposed to humanoid collectives of the 4th efflorescence all of the major characteristics of these isolated groups, including their peaceful and collectively non-aggressive characteristics, are quickly assimilated to collective modes characteristic of the humanoid collective in the 4th efflorescence. In this process which has been unfolding, as it were, in front of our eyes during this century, we may reconstruct the nature of the general trend which has been going on since the end of the upper Pleistocene somewhere between 15,000-20,000 years ago. Once a humanoid collective is firmly established in one efflorescence, exposure to the next or contact with humanoid collectives who have already attained the next stage, will quickly propel the earlier efflorescence into the later one.

The presupposition is that if, for some reason, all the collectives of the 4th efflorescence had been annihilated before making contact with the isolated collectives still established in the 3rd, those collectives would in time, independently and by themselves have evolved into the 4th efflorescence phase. The critical factor is of course time, the time it takes for the institutionalization of superego and the mother figure to become customary, rather than a new mode of learning behaviour attained with difficulty by each generation. Thus the point at which 3rd efflorescence society transmuted itself into 4th would seem to be an 'accidental' rather than a necessary evolutionary development, since there is no way of predicting at what point in time this would occur. Only because we have witnessed such transmutation in modern times does it seem to be 'necessary' rather than accidental, although from this point of view some impelling force or catastrophe has clearly been at work. At any rate, it can be said that society which has reached the stage of 3rd efflorescence possesses the potential for attaining the 4th efflorescence stage.

Needless to say the transition from the 3rd to the 4th efflorescence of humanoid development is often neither painless nor constructive, as can be seen from the innumerable case studies done on the process of 'de-tribalization' Africa and in other parts of the world. If not carefully nurtured and educated into new behavioural modes by the group of the 4th efflorescence who comes in contact with the exposed 3rd efflorescence group, usually as a colonist if not as a conquerer, the result is often calamity for that group. Psychologically, individual ego-recognition and libido aggression among members of a collective of the 3rd efflorescence are muted, while the collective ego recognition and libido aggression that allows the group as a whole to protect its individual members is yet unborn. The chronicles of modern man are filled with instances of American indians drinking themselves to death on 'protected' reservations out of despair and helplessness, and landless Africans who have lost their grazing lands starving to death in innocence and wonderment on the edges of industrially developed enclaves of towns and cities.

In parts of southern Asia and Africa where ideologies of collective behaviour have been superimposed on kinship, art, religion and witchcraft to manipulate individuals to action, even more grisly results have been seen where, in the total collapse of the achievements of humanoid collectives in the 3rd efflorescence, individual ego-recognition and libido aggression are released with primordial force, and a total reversion to conditions of the lower Pleistocene result.

And in all of this there are instances which represent intermediate stages, typical of the 'Third' or 'underdeveloped' world, especially in East Asia, central and southwest Asia, India and parts of North Africa and Asia Minor. For the count is not yet in on the 4th efflorescence of humanoid development, and whatever part of the world we consider, whatever segment of present day existing human collectives, at the very best human development is still in a testing stage of the 4th efflorescence where the tally between the late stages of the 3rd efflorescence of human development and the testing stages of the 4th efflorescence remain indeterminate.

In terms of the general trends of humanoid evolution, the 4th efflorescence of human development will not have fully blossomed until we can see the direction into which it is evolving in a 5th stage of human efflorescence, and that direction most surely can only be surmised and not yet predicted with any sense of certainty. Thus, if the nations of the 'developed' world of Europe, North America and Japan represent the testing stage of the 4th efflorescence, while those nations of the 'developing' world represent the late stages of the 3rd efflorescence, the true disparity between these blurred phases are difficult to judge. While over a shorter period of time it would appear that nations of the 'underdeveloped' world have emerged too slowly into the 4th efflorescence stage, over a longer period it may prove that nations of the 'developed' world have emerged too quickly. While from the point of view of the short term, it appears that the nature of the 'developing' world are at an adaptive disadvantage vis-à-vis the nations of the 'developed' world, over a longer period of time it might appear that nations of the 'developed' world have lept too rapidly into the 4th efflorescence, abdicating advantageously adaptive features of the 3rd efflorescence that will in the end make them maladaptive in the 4th efflorescence.

The mindless and avaricious consumption of natural resources, and the long history of genocidal war and material and cultural self-destruction on the part of the 'developed' nations at least suggests that the latter might in the end prove to be the case. This however is a difficult matter to judge from our limited historical perspective, for it might prove that no matter how gradual the transition from the 3rd efflorescence to the 4th may be, the crisis of the ascendancy of collective ego-recognition and libido aggression is so formidable, whenever and however it occurs, that no vestige of advantageous adaption from the 3rd efflorescence will be adequate to ameliorate it. While, obviously, if the humanoid species is to survive, evolution must continue into a 5th efflorescence, which will of necessity serve to ameliorate the crisis of the fourth, and it may be that what appears to be advantageously adaptive about 'underdeveloped' nations still in the latter stages of the 3rd efflorescence is that there are aspects of the particular way in which they are retarded in their development into the 4th efflorescence which may presage the evolutionary solutions of the 5th.

What seems most striking about the 'underdeveloped' nations and regions of the world that differs from the 'developed' nations is the existence of wide-spread and often fanatically held religious philosophies which appear to retard a more rapid transition into fully 'developed' nations. These religious philosophies such as Taoism, Confucianism, Buddhism, Hinduism, and even to some extent Islam and Judaism, have in common ego-negating and libido-abdicating values. Often, even when collective ideologies such as Communism are transplanted to 'underdeveloped' regions they take on this same sense of ego-negation and libido-abdication, at least where ordinary members of the common collective are concerned. While Christianity may also be seen to contain many of these same value affects, in countries where these values are fully developed in Christian practices in orthodox Catholicism, these countries remain on the fringes of underdevelopment. Only where Christianity has for the most part been abandoned or else replaced with a form of Protestantism which emphasizes ego-recognition vis-á-vis God's acknowledged grace and libido aggression in performing the actions that prove this ego-recognition as 'election' do we find the countries and regions that are generally considered to be the most 'developed.'

In Communist countries considered 'developed,' Christianity has been abandoned for an atheism which gives free reign to both individual and collective ego-recognition and libido aggression. 'Developed' in this sense implies countries which possess the greatest amount of material power, both economic and military, and where individuals of those countries are most easily, willingly and customarily organized to produce the wealth which benefits not so much the individual, but the country as a whole in terms of its power, ie., its ego-recognition or prestige and its libido aggression or its willingness to commit acts of hostility or aggression to protect that prestige. Where in 'underdeveloped' countries or on the fringes of the 'underdeveloped' countries individuals consider it a priority to negate their own individual egos and abdicate their individual libido aggression, there is little collective affect available for the purposes central to attaining the status of a 'developed' country.

In a fully 'developed' country, superego is taken for granted by the members of the collective; that is, it is there, it exists in the psyche of the individual, but the individual does not dwell on it or devote much affective energy to it, as it exists more as a matter of custom than a conscious or engaging value. Thus, there is a free flow of affective energy between individual and collective ego-recognition and libido aggression. The one is easily transformed into the other, and the two points of ego-recognition and libido aggression, individual and collective, act as checks and balances on each other. Where in 'underdeveloped' countries ego-negation and libido abdication is held as a central individual value, this free flow between individual and collective ego-recognition and libido aggression is of necessity hampered if not dammed up altogether.

Thus the most notorious tyrants of modern times, Stalin and his adept pupil, Hitler, attempted to make their nations more 'developed' and more 'powerful' by using various devices to de-institutionalize superego among the collective. Stalin used a form of Communism as a counter philosophy to inculcate among his people the idea that the collective historical teachings which embodied superego, such as religion and classical philosophies and their expression in literature and art, were false. By inculcating the idea, as well, through example of force and political repression, that individual superego was also inimical to this counter-philosophy of negative superego, he attempted to unleash naked ego-recognition and libido aggression in the service of the collective ego-recognition and libido aggression. However, since this counter-philosophy, as an expression, nonetheless, of the 4th efflorescence, was based on the ability of individuals to sublimate individual ego recognition and libido aggression in the collective one, this meant that the focal point of such sublimated ego and libido became the single individual leader, i.e., himself. In such circumstances, where superego is all but eliminated and the leader becomes the focal point of sublimation, which is an outgrowth of institutionalized superego in the first place, the leader then hovers between being deified, on the one hand, and being, on the other hand, a target of all the unleashed individual egos and libido which ultimately strive to replace the leader as an overwhelming object of ego-recognition and libido aggression with their own.

The state or collective veers frantically between enthralling an enormous unleashing of affective power to its own ends, and ultimate chaos and self-destruction. In the case of Hitler, a particular group, the German Jews, who represented an unusually fertile source of superego expression both in their collective philosophy and in the number and quality of individuals who represented superego in the particular pursuits they engaged in, such as medicine, priesthood, art, philosophy, were made the target of super-ego negation by the German nation as a whole. Even though individual superego among members of the 'pure' German collective were relatively untouched, there was a concerted attempt to eliminate

Christianity and all of its associated philosophies and values from the collective superego, and it became a collectively reinforced value of the superego to express ego and libido through force in the elimination of all of those individuals and cultural artifacts which represented selected aspects of superego.

In modern Communist China, Mao also came under Stalin's influence. The visible repository of Chinese superego in the works and artifacts of classical Chinese philosophies and religions were destroyed, and all individuals in the collective who represented, as individuals, superego attachment, such as doctors, lawyers, artists, writers and professionals and intellectuals, of every kind were eliminated, an affective value of collective superego being placed on the destruction of all these representatives of superego. Again, as in the case of Stalin and Hitler, where individual ego and libido were unleashed in circumstances where superego was already firmly institutionalized, the leader, here Mao, became the sole representative of the superego he strove to eliminate in all cases but the objectification of his own person by the masses. Again, as in the case of Stalin and Hitler, where the deified leader demonstrated his mortality by dying, his anti-superego philosophy died with him, and the society reverted normally to customary acceptance of representative devices of superego in all of its social and behavioural modes.

Where ego-negating and libido abdicating philosophies persist in the 'underdeveloped' world of both Communist and non-Communist countries, the process of sublimation of ego and libido takes a different direction from that in developed countries. In developed countries in which these philosophies are absent, sublimation of individual ego and libido is transferred almost immediately to collective ego-recognition and libido aggression. In those countries and regions that are in the late stages of the 3rd efflorescence rather than the early stages of the 4th, ego and libido are also sublimated, but into diffuse forms of their negatives which in no way eliminate ego-recognition and libido aggression as drives but merely displace them, delaying the effect of their ultimate force when sublimated into collective ego and libido proper. The ego-negation and libido abdication philosophies such as Taoism, Buddhism, Confucianism and Hinduism use a number of psychological and psycho-linguistic devices to teach that once ego is recognized, it can be negated, and that once libido is expressed it can be abdicated. The ultimate objective in these philosophies is to replace ego with a non-ego, libido with anti-libido. From the point of view of Freudian psychology, non-ego is the same as rational consciousness, and anti-libido is of course superego. But there are vast and critical differences in the way that these psychological concepts are understood in the ego-negating and libido abdicating philosophies of the Far and Middle East, and the way they are understood by psychologists in the developed countries.

From the point of view of modern psychology, when ego is negated and rational consciousness is achieved, ego does not disappear: it becomes customary, taken for granted by the individual, and is robbed of extraneous affect from libido or superego which otherwise makes existence a constant source of anxiety. Similarly, from the point of view of modern psychology where libido is abdicated, it by no means disappears. The aspects of libido which interfere with rational consciousness, thereby disturbing the natural flow of ego affect are sublimated into superego, that is, controlled by conscience, but libido as such, in its 'normal' aggressive and hostile flow of affect, far from being eliminated, is supposed to be, under these ameliorating conditions, more free to express itself than before. However, in ego-negating and libido abdicating philosophies, ego is supposed to be entirely eliminated; it is to be replaced in part by rational consciousness, known as 'enlightenment' in these philosophies. The affective part of ego however which cannot in actuality really be eliminated is displaced into non-ego, which is only partly a term for rational consciousness. As 'enlightenment,' 'non-ego,' like 'anti-libido' has no specific locale; it is neither entirely within the self nor entirely outside of the self. It is, as it were, a cosmic consciousness, a primal force which is in all things, internal and external, and is viewed as a first cause of all creation.

In certain of these religions, this non-ego as a primal force in all things is keyed, in locale, to a semi-mythical agent of the primal force, such as a Christ or as a Buddha, whose life and whose teachings are more often than not used as a device in the way of an example to assist in teaching the noviate the way of ego-negation and libido abdication. In these kinds of religious philosophies the Christ or Buddha figure or the other agents of sublimation utilized are not viewed as a person or human or such, or even as an objectified God figure. He is in fact basically irrelevant to the philosophy as such, being used as a marker or guidepost, a device, by which ego-negation and libido abdication can be more easily achieved.

The ego which is negated becomes identified with a non-ego that encompasses the universe, and is, from a modern psychological point of view, a massive form of egotism, albeit in a negative way. The same is true for anti-libido, for the cosmic superego which displaces libido in these religions is the force which controls, moves, makes, destroys, creates all things in the cosmos, and as such is a massive form of libido expression, albeit again in a negative, or perhaps we should say, positive way. This anti-libido or cosmic superego places pejorative values on any individual or even collective form of libido expression, making the individual or the collective entirely pacificist. The affect of libido when it becomes anti-libido or cosmic superego is expressed by activity which is seen to be in harmony with the primal force, and this usually involves mandated action that is benevolent, kind, self-sacrificing, self-abnegating, or entirely passive. Thus, the affect of libido is utilized in the reverse of its normal modes of aggression, violence, hostility and sexual activity. Even when anti-libido is manifested by extreme passivity, the affect of libido is utilized in a reverse form, that is, to control and direct itself to the reverse of its natural impulses.

The difference between these religious philosophies and Roman Catholicism, for example, is another hallmark, along with collective ego recognition and libido aggression, of the difference between the late stages of the 3rd efflorescence of humanoid development and the testing stages of the 4th. In Roman Catholicism, the locale of the primal force is located with precision: it is outside of the individual ego, and of course, outside of the collective ego as well. There is no possibility of confusing or identifying individual ego or collective ego with the non-ego of the primal force, or God. There is no way to conjure or identify individual or collective libido with a cosmic anti-libido or superego which, benevolent or otherwise, has the power to create and destroy according to its own divine precepts. The Christ in Roman Catholicism is precisely defined in terms of his ontology, his birth, locale of life and time, and manner of death, and he is not viewed merely as a symbol or representative of the primal force, but as a humanoid-like individual with specific powers, privileges and characteristics in relation to both the primal force, or God, and individual and collective human beings. It is this specificity of locale regarding the separation of the actions, affect, and ego of the individual human and the collectivity to which he belongs from the action, affect and ego (or anti-ego) of the cosmos, which is a critical hallmark of humanoid development in the 4th efflorescence.

Specificity of locale of action is an indication that while superego is institutionalized it is no longer a matter of affective or emotional engagement of humanoids in the 4th efflorescence: it is taken for granted much as the use of fire and the making of tools became taken for granted when man evolved from lower to upper Pleistocene. Nonetheless, given the calamitous results of humanoid evolution in its 4th efflorescence, the question remains, has mankind come to take for granted the institutionalization of superego too quickly in his development? The casualness with which mankind has come to take institutionalization of superego for granted in the 4th efflorescence of his development is reminiscent of the casualness which he took for granted in individual ego-recognition of the lower Pleistocene in the 2nd efflorescence of his development. The casual acceptance of ego-recognition and libido expression as humanoid institutions in the lower Pleistocene wrought such havoc on human collectivities that the very survival of the race remained in question for 400,000 years. With this casual acceptance of ego-recognition and libido expression, mankind stood precariously on the brink of self-destruction for all of this time, and not until superego developed and became institutionalized in the upper Pleistocene did humanoid evolution once again display the happy prognosis it had obtained during the first efflorescence of humanoid evolution 1-2 million years ago.

Undeniably, the development and institutionalization of superego in the 3rd efflorescence changed the phenomenological aspect of humanoid evolution to such an extent that, compared to humanoids of the 3rd efflorescence, humanoids of the 2nd appear almost as 'creatures from another planet,' a sub-special differentiation hardly recognizable as 'human.' Indeed, when explorers and adventurers of the 'developed' world in the 4th efflorescence first discovered the indigenous inhabitants of geographically isolated regions who were firmly still entrenched in the 3rd efflorescence, they were seen as 'primitive' and 'savage.' However, among the major civilizations of Asia and Africa where religious philosophies of ego-negation and libido abdication were maintained, the reaction of individuals from the 4th efflorescence to these was, and remains, a puzzling one. The unique achievements and the persistence of these civilizations causes humanoids of the 4th efflorescence to wonder whether or not they left superego and consciousness behind them before they really understood them.

From the point of view of modern psychological analysis in an evolutionary framework, we know that the development and institutionalization of superego had far reaching social, cultural and psychological implications which are responsible for modes of behaviour that continue to make human collectives even in the 4th efflorescence viable. It is the further developments of humanoid evolution in the 4th efflorescence in fact which threaten this viability which has allowed humanoid development to flourish since the upper Pleistocene. Objectification of all phenomena, natural, social, cultural and psychic has allowed humanoid collectivities to evolve into a 4th efflorescence, but this objectification is based, both from a psychological and evolutionary point of view, on the subjective experience of natural phenomenon, and in particular the subjective experience, which we now objectify, of ego, libido, superego and consciousness. The evolutionary pathway by which subjectification of phenomenon changed to objectification is a subject for further evolutionary analysis, and as with all evolutionary changes, the specific pathway by which it occurred was no doubt a matter of 'accident.'

In the need for the proliferation of the institutionalization of superego, a more rapid method was needed to inculcate in the young of the collective and among outsiders who joined the collective through marriage or other means, the devices of institutionalization. For these purposes, subjectification of phenomenon, and the means of expressing and teaching it, possessed a vehicle of communication that was slow and cumbersome and could not be standardized or regulated. A pictorial form of language which possessed in a single character multitudinous meanings on a variety of phenomenological levels was insufficient to the needs of rapid promulgation of the devices of superego and its institutionalization. Written language and spoken modes of communication were thus developed where words and characters had more less single meanings, rather than multitudinous ones. Devices of superego institutionalization that were previously expressed in a single symbol or word were altered to be expressed, when spoken, in a long chain of words each with a more or less single meaning, or when written, to a long chain of signs each with limited symbolic import. This alteration in the mode of verbal and written communication resulted in a transition from the subjectification of phenomenon to its objectification.

Naturally, as in all evolutionary processes, this change in the method of communication evolved gradually, and the degree of objectification varied directly with the degree of change in the modes of spoken and written communication. By the time of the Phoenicians and the early Greeks, some 4000 years ago, this alteration had become so complete that single written signs, as letters, no longer carried any intrinsic symbolic significance at all. Letters which were signs devoid of the symbolic meaning they had once had in their middle Eastern linguistic antecedents, had to be grouped together to form a single word whose meaning was *de facto* limited to specifics. The words composed of letters reflected in phonetic sounds the spoken language that had been developed among these peoples, where declensions of verbs, conjugations of nouns, definition of gender and person, connectives and prepositions all ensured the specificity of meaning. A single word was no longer able to contain a plethora of symbolic meanings as it had in its antecedent tongues, for specific markers accompanied it and changed its form according to the particular situation and phenomenological level it was meant to designate. Objectification of phenomena had occurred, and methods which were meant to inculcate the devices of superego and its institutionalization allowed these devices to be entirely removed from superego and to be utilized once again for the purposes of ego recognition and libido aggression.

When subjectification of phenomena was displaced by objectification of phenomena in this way, the individual was removed from the subjective experience of superego which had allowed him to integrate sense of self with the sense of the welfare of the collective as a whole. But objectification of phenomena through specific modes of thought and communication also displaced mankind from a sense of mere individual ego recognition and libido aggression. Objectification of phenomena did not hurl man back to the level of development he had known in the lower Pleistocene. On the contrary, his sense of ego recognition and libido aggression was identified or associated with the collective, as objectification of phenomena was an absolute way of removing the individual from any experiential sense of himself. Thus, in this rapid leap into the 4th efflorescence of humanoid development, mankind attained an objectification of phenomena which led in effect to a denial of the experiential sense of self. What those civilizations which remain in the late stages of the 3rd efflorescence have to tell us in the 4th efflorescence in such a compelling way is that denial of self is the ultimate illusion. Self may be abdicated, but it cannot be denied, for ego-recognition and libido aggression are the *sine qua non* of human evolutionary development, the criteria of humanoid existence as such. If we delude ourselves into believing that self can be denied, it will only be reasserted in a more virulent and ever more uncontrollable form. On the other hand, ego-negation and libido abdication are firmly based on the acknowledgement of self. For egotism to be contained it must first be acknowledged.

Chapter IV

FAR EASTERN PHILOSOPHIES ATTEMPT TO TAP THE POWER OF THE SUPEREGO

The evolutionary development which led man from the beginning of the upper Pleistocene through the entrenchment of the 3rd efflorescence to its present late stages of the 3rd efflorescence and the onset of the 4th was in the range of 100,000 years. As the well-spring of man's superego was tapped in isolated geographical regions, in isolated collectives and under differing conditions of erratic social and psychic development, the varieties of culture produced in this 100,000 year period were no doubt innumerable. From the myths and legends of cultures that have evolved from these Pleistocene antecedents and are today extant, we can only guess at the number of distinct cultures that must have flourished in this period of time, and disappeared, either because they evolved into an other more adaptively advantageous forms or were absorbed by other cultures. Absorption of one culture by another would have in some cases been voluntary, where the culture of one group was viewed as possessing advantageously adaptive features that could be easily assimilated to collective modes already well-established in that collective. In other cases, the institution of slavery inherited from the lower Pleistocene and which in fact made development in the 3rd efflorescence in the upper Pleistocene possible, was responsible for the obliteration of cultures with less adaptively advantageous features. Where the vestiges of individual ego-recognition and libido aggression from the lower Pleistocene had reached homeostasis with collective modes of the 3rd efflorescence, under conditions of a slave society, social leadership, war, and trade were all possible, albeit in perhaps vaguely defined forms, at least by standards of the late stages of the 3rd efflorescence or the 4th.

Thus, in these cases outright conquest and subjugation of one collective with a distinct culture by another was possible, and the fate of the conquered culture ranged in possibility from genocide, to organized enslavement, to peaceful assimilation of the conquered culture, depending on the exact set of circumstances, and the differential in culture and technology between the two groups. The point is that by the time of the entrenchment of the 3rd efflorescence, say, about in the mid-point of time between upper Pleistocene and post-historical time, about 50,000 years ago, all the features that we today think of as characteristic of civilization as we know it in the earlier stages of the 4th efflorescence or the later stages of the 3rd, were already present in the collective modes of humanoid efflorescence. Once the psychic well-springs of the superego were tapped and became institutionalized, spoken language, written language, art, science, technology, law, military organization and protocol, and governmental forms were already present, although by no means were these forms of civilization standardized from one collective culture to another, or necessarily all equally well developed in any given collective culture. The fact that no actual archaeological sites have been found of the remains of cultures that possessed all these sophisticated attributes of civilization in no way proves that they did not exist.

Until a few years ago, the oldest post-historical sites known were of the ancient Egyptians, and only recently even the advanced methods of carbon dating and other scientific means to date archaeological sites with exactitude have given a more precise date for these remnants of Egyptian post-historical civilization. Before these scientific methods were developed for archaeological dating, Egyptian tombs and pyramids were thought to be no more than about 5000 years old, while now we know the earliest of them are from 12,000 to 20,000 years old. Previously as well, there were thought to be no civilizations elsewhere older than about 3,000 years old, but recent excavations in southwest Asia, India and Turkey have revealed the remains of cities as old as 10,000 years, cities which appear to differ but little from more recent remnants of civilized society of 2,000-3,000 years old. The fact is of course that the building materials that humanoids have used to construct their cities and burial places are finite materials, and given enough time even the vestiges of these building materials will disappear. In addition, as in the case of sites found in southwest Asia and India, as well as in almost every other part of the world where civilization has had a continuous pathway of efflorescence, cities are built one on top of the other. Where the earliest sites have been discovered, the remains of the earlier cities have been found under the rubble of a continuous series of later ones. Improved methods of scientific archaeology will no doubt in the future enable archaeologists to pinpoint the remnants of civilization even earlier in time than those know today. Certainly however from archaeological evidence and the general trend of humanoid evolution, there is a finite limit to the time period in which remnants of civilization as we understand it can be found, and that limit is probably approximately 50,000 years ago, somewhere halfway between the upper Pleistocene and the 4th efflorescence in which time we are now living.

It is typical of man's *hubris* in the 4th efflorescence to believe that only in his own period of historical development have the accoutrements of civilization been developed. Racism, prejudice, and national chauvinism which is also typical of mankind in the 4th efflorescence, where free-ranging ego-recognition and libido aggression are so closely identified with collective ego and libido, make it difficult for individuals in the 'developed' world to recognize achievements of those in the 'underdeveloped' world. Liberalism, which is a confluence of many factors, including vestiges of more orthodox Christianity, secular nationalism, and an inadequate attempt to emulate the ego-negating and libido abdicating philosophies of 'underdeveloped' countries, has done little to ameliorate this situation, and the human heritage of 'underdeveloped' countries and primitive societies of the 3rd efflorescence are given no more than mere token respect. Science, no doubt because of the terrific boost it and its handmaiden, technology, give to the collective ego and libido of man in the 4th efflorescence, is almost universally believed, in 'developed' countries, to be an entirely new invention of civilization in modern times, the criteria of progress, the source of wealth and the pathway of future evolutionary development. In fact the obsession with science and the pride of its possession is a definitive feature of the superiority complex that has devolved around modern man's collective ego in the 4th efflorescence. The modern hysteria that surrounds the production of more and more scientifically precise weapons of war, as symbols of collective power and strength, would be an obvious representation of a manic and unrestrained 'penis pride' to any Freudian psychologist who would venture to make the connection between the individual ego and libido and the collective one, which has become the basis for our present modes of social and cultural behaviour.

Nonetheless, all contemporary hysteria and obsessive fixations to the contrary, science is not the newest invention of humanoid development, but the oldest. In the second efflorescence of humanoid development during the lower Pleistocene 500,000 years ago, it was the discovery of science that followed hard upon the first raw glimmerings of ego-recognition and libido aggression, making the customary use of fire and the regular production of intricate tools, mostly weapons, possible, in this earliest, most hazardous phase of humanoid efflorescence. The tapping of the psychic potential of the superego during the 3rd efflorescence of humanoid development allowed science to develop to a state of sophistication that is almost unimaginable even by modern standards, although at least in the mid-point of the 3rd efflorescence, due to the latency of both individual and collective ego and libido, science was not used for the purposes of war or aggression. In this period, magic, art, religion and medicine emerged in a synergistic whole with science and resulted in the creation of the infinite assortment of datum we are used to taking for granted as the accoutrements of civilization. The sciences of agriculture, herding, domestication of animals, and their breeding, cooking, wine production, architecture, pharmacy, city planning, road-making, map making, systems of writing, arithmetic, weather forecasting, systems of heating and plumbing, ship building, and surgery are among the plethora of inventions produced by science in this period.

From this perspective the sciences of producing tin, bronze, iron and pottery are merely late spin-offs from the achievements of scientific discovery in the 3rd efflorescence, and other sciences which we can only speculate upon because their exact nature has been lost in the confusing annals of post-historical time were no doubt also developed, to perhaps be created afresh in our own modern times. Thus, the almost universally held notion of our modern times that science *per se* is only a few hundred years old, derived from the specific philosophical discoveries of the Greeks less than 2500 years ago from a continuous line of deductive reasoning and trial and error experimentation peculiar only to so-called Western man as he developed in the southern and northern parts of the West European continent and in North America is entirely fallacious and grieviously misleading, playing as it does upon the delusions of grandeur so frenetically and self-destructively held by 'developed' countries in the 4th efflorescence.

For anyone who has had direct experience with the rigours and aesthetics of pure scientific thought, the subjectivity of the experience it both releases and demands is not easily forgotten. Mathematical formulae and the abstract thought it represents is a symbolic mode of thought with which even the finest poetry pales by comparison. Like music, mathematical thought exists in a plane outside of the verbal or written objectification of phenomena, and it is expressed in symbols which harken back to a time when all thought was a matter of the subjectification of phenomena. While mathematical symbols and formulae like musical notes are not perceived in our time as having multitudinous levels of meaning, increasing agility and success with mathematical development is beginning to raise just this kind of conundrum. The mathematics of Einstein, for example, has raised questions of both a philosophical, religious and even a social nature which are of no interest to the pure mathematician as such, but come into the purview of all people, implying as it does significant insights into their lives on many levels at once.

Einstein, like many modern mathemeticians who faced the same problem with their mathematical work as he did with his, such as Heisenberg, Planck, Russell and Whitehead, to name just a few, felt compelled to expostulate on the implications of the findings of his mathematical and physical research and to branch out in his professional capacity as a social and religious philosopher. The subsequent mathematical and physical work or branches of physics and mathematics opened up by Einstein, Heisenberg, Planck, Fermi and others, in the fields of nuclear fission and fusion are far more important today in terms of the social and political polemics and modes of collective behaviour that they have produced than the purely scientific aspect of this work ever was. Mathematical symbols and formulae in fact are quite closer in substance to the ancient hieroglyphics of early post-historical civilizations such as the Egyptians than they are to the objectified word chains that are used as the sole mode of communication in the 'developed' countries in the 4th efflorescence, and to a large extent by 'underdeveloped' countries in the late stages of the 3rd efflorescence. This suggests the ultimate logical deduction that modes of verbal and written communication in civilizations prior to these were very much akin to what we now think of as 'mathematical' expression. It is a stupendous revelation to recognize that well-developed civilizations of the 3rd efflorescence communicated in terms of mathematical symbols, for it allows us to perceive finally the magnitude of their scientific orientation. Thus, even though science developed originally in the 2nd efflorescence of humanoid evolution, before the development and institutionalization of the superego, for reasons that are tied to both the psychological and biochemical aspects of superego development, science flourished and grew to its present status only during the 3rd efflorescence of humanoid development.

Superego developed as a result of the increasing prominence of the female in general and the mother in particular during the 3rd efflorescence in the upper Pleistocene. The institutionalization of slavery during this period freed the female from the ravages of unrestrained ego-recognition and libido aggression on the part of the male, and longer and more secure periods of child-rearing allowed male offspring to develop closer relationships with their mothers which were internalized in maturity in the psychic component of superego. Longer and more secure periods of childhood for male offspring permitted a delay in the infusion of male hormonal systemics in their neural biochemistry, resulting in a balance in male neural biochemistry between male and female hormones, or between hormones of the lower, reactive and upper, cognitive brain systems. Whereas from the biochemical point of view the development of science in the lower Pleistocene was associated with the dominant hormonal systemics of the lower brain system, or the automatic nervous system, in the upper Pleistocene this development was mightily augmented by increased profusion of hormonal systems associated with the upper brain, or voluntary nervous system. The blend of hormonal infusions from these two systems meant that scientific awareness was equally associated with both instinct and cognition.

In the first flowering of scientific development during the 3rd efflorescence in fact there was no area of human experience, sensation or proclivity that properly lay outside of potential scientific mastery. All experiential activity, from sensitivity to bodily reactions at an involuntary level, to awareness of all aspects of the external cosmos, lay within the scope and purview of scientific interpretation, conceptualization and ultimate control. The sensations of bodily rhythms of an involuntary nature, such as the reactions of the nervous systems, muscle systems and all bodily organ systems were revealed to conscious awareness through the fusion of upper and lower brain hormonal systemics, and could be used as a model, in the cognitive brain, for the workings of the external universe. With the subjectification of phenomena as the sole mode of thought and communication there was no dividing line between magic, science, art, and religion. Under optimum conditions the powers of at least certain individuals to manipulate their own organic and psychic systems, the organic and psychic systems of others, and even aspects of the external environment, would have appeared, from the perspective of later stages of humanoid development, to be almost God-like.

Even today among so-called 'primitive' societies of the 3rd efflorescence that still exist in Africa, central Asia and among American Indians of North and South America, the vestiges of these ancient abilities are still known to exist, as the detailed chronicles of the American anthropologist, Carlos Castenados, of his encounters with tribal magicians of Mexican Indian tribes reveal. With the aid of the unique pharmacopoeia of these Indians, which include the hallucinogenic muschrooms and other roots found in the southwestern United States and Mexico and whose biochemical effects are to blur the division between the neural biochemistry of the upper and lower brain systems, the magicians of the tribe possess the extraordinary powers to manipulate their own minds, and the minds of others who may or may not be also under the influence of these drugs, and even apparantly certain aspects of the external environment. Among other vestiges of societies in the middle phase of 3rd efflorescence that still exist in the so-called 'developed' world of Europe, as among the Druidic sects of the British Isles and in Central Asia, similar kinds of extraordinary manipulation of self and the environment are also known to occur, and for central Asia these kinds of phenomena have been documented in the book, *Psychic Discoveries Behind the Iron Curtain*, where certain individuals in this region still possess considerable powers of both controlled telepathy and telekinesis.

Even since the development of Western science and mathematics, the great discoveries in these fields have been associated with religious and transcendental philosophies. Pythagoras, the inventor of modern geometry, was associated with a religious cult which was believed to have had its origins in ancient Egypt, and the first theories of physics developed by Democritus and Lucretius were as much concerned with transcendental metaphysics as physics proper. The founder of modern chemistry, Mendeleyev, was engrossed in the metaphysical and transcendental philosophies of central Asia and China, while Newton's obsession with transcendental metaphysics and spiritualism has long been an embarrassment to a modern scientific tradition which has held the egotistic notion that science is self-consciously and rationally created through the objectification of phenomena. Pascal and Einstein were two mathematicians who publically proclaimed on numerous occasions that their own religious faith and the notions of order that this faith imbued them with were integral to their mathematical notions of universal order. In fact great mathematicians and truly creative physicists who have been atheistically rather than religiously inclined have been the exception rather than the rule, while in a case such as Descartes his mathematical creations were integrated with a philosophical view of the universe that was rationalistic to the point that it verged on Platonic transcendentalism.

In more modern mathematicians such as Russell, Whitehead, Quine and Von Neuman whose view of the natural world was more or less atheistic their works have, willingly or unwillingly, been focused on mathematics as semantics, or forms of code or communication, and whose implications, in the synthesis of all the moderns arts and technologies of our times, are only beginning to be understood. It was only Carl Jung who understood in his scientific work that religion, myth, and synthesizing modes of language and communication were not only the object of science but also its subject. Religion, myth and modes of communication were not only the means by which historically science has come to be understood, they were the very subject of science–science was the stuff of which man's mind was made, in the guise of religion, myth and communication modes.

In those 'underdeveloped' countries which have not lept as rapidly into the 4th efflorescence as the 'developed' countries, we find philosophies and practices that synergize, thanks to the tapping of superego potential, all of man's cultural modes harnessed to the ends of the collective ego and libido, such as it is in these societies. Philosophies of Taoism, Confucianism, and Buddhism have resulted in the creation of exotic arts, sciences, and technologies, the trained practitioners of which have at various times been used by the military or bureaucratic organs of countries in Asia for purposes that would further the ends of the state. Temple schools in these regions have produced entire classes of warriors and administrators whose physical and mental skills were, and are, extraordinary compared to their counterparts in the 'developed' world. While the 'Boxer rebellion' of China in the early part of this century demonstrated that even the extraordinary physical and mental powers produced by training in these arts and sciences so closely associated with transcendental and supernatural philosophies of Asia, were helpless against the bullets and guns of the 'developed' world, nonetheless, in hand to hand combat the 'martial arts' training of the East has proven so advantageous that today it is a regular part of training and physical exercise in every army and police force in the world. As well, the particular mental abilities of people trained in these philosophies and their resultant practices and their potential use in the service of the collective is a serious object of study in 'developed' countries, although the intricacies of the training remain rather impervious, due to their subjective nature, to any understanding of them which is expressed through objectification of phenomena.

The text of *Tao* begins, 'The *Tao* is that which cannot be named. That which can be named is not the *Tao*.' Primal science, like religion, is based on paradox, and paradox, by definition, is that which does not lend itself to a total objectification of phenomena. 'Give up learning, and put an end to your troubles,' the texts of *Tao* explain, for in those civilizations which have hesitated before entering into the 4th efflorescence, there is extreme scepticism about the utility of the total objectification of phenomena. Not that there is no objectification of phenomena, but in these civilizations to integrate subjectification of phenomena with objectification, and not to lose the peculiar strengths gained by the individual when the two are combined in a proper balance is the aim. In the special training of these elite cadres in such civilizations, where the well-springs of superego are tapped for the utilization of consciousness, meditation on paradox is combined with special physical excercises that were developed, in time, as an adjunct to these meditations. The exercise usually begin with 'standing on the horse,' learning to stand for long periods in a semi-crouched position, halfway between four-footed and two-footed posture. It is believed in these philosophies that the greatest strength of a man resides in the muscles of the entrails, and if this strength, through exercises of balance, coordination and movement, can be displaced to other parts of the body, extraordinary strength can be attained.

From both a psychological, physiological and biochemical point of view what these methodologies attempt is an integration of the lower and upper nervous systems. From a psychological point of view, intuition and cognition are integrated, and the emotional flow from the aggresive aspect of libido and ego are infused with superego and consciousness, while the cognitive aspects of superego and consciousness are infused, in a feed-back system, into ego and libido. From a physiological point of view, the extreme reactivity of the lower nervous system is infused with the cognitive facilities of the upper nervous system and vice versa. As from the point of view of evolutionary development, the entrails comprise the primal nervous organ and is the actual origin of all differentiated parts of the nervous system, there is particular attention payed to developing a 'consciousness' of the activity of this part of the involuntary nervous system. All parts of the involuntary nervous system are conditioned, through exercises and meditation, to 'think,' to be conscious, while the voluntary nervous system, the upper brain, is conditioned to be aware of the reactivity of the involuntary system.

The upper brain becomes a kind of clearing house for information from every part of the involuntary system, so that communication between various parts of this involuntary system are able to communicate with each other. As the purpose of these exercises and meditations is usually, although not always, for combat of some sort or another, usually physical, the strength of powerful parts of the involuntary nervous system, for example, the bowel, can be displaced to weaker parts, for example, the arm, the hand, or even the fingers. Conversely, through this 'clearing house' system of physiological intra-communication, the individual can learn to see, hear or think, i.e., cogitate with organs that are largely directed by the involuntary nervous system, such as the skin, or the muscles of any given part of the body. In this way, these philosophies and practices of civilizations in the late stages of the 3rd efflorescence allow a rationalization, or objectification, of the witchcraft and magic of early stages of the 3rd efflorescence, and direct these practices to objectives that serve the purposes of the collective, rather than just the individual.

These philosophies and practices can also be explained from a biochemical point of view. The major biochemical or hormone active in the lower nervous system, acetycholine, is a compound with two nitrogens in its complex that contain two points of information, containing thus the elements of reactivity rather than cognition *per se*. These two points of information may represent the co-existing potential, say, for both flight and attack, or ingestion and excretion, depending on the purpose of the organ where these compounds are located. As a system of information, this compound represents a non-binary system: it is the 'name that cannot be named,' the phenomena of paradox. This non-binary or paradoxical system allows for homeostasis or equilibrium of the organ, while at the same time it contains the potential for a certain decision, i.e., flight or attack. The transition from the system of flight *and* attack to flight *or* attack is triggered by the hormone of the upper nervous system, norepinephrine, which contains a single nitrogen atom in the complex as a single point of information, which makes of it a binary system of information. In order for the upper brain to act as a clearing house for the information of the lower brain, the binary system of information in the upper brain must be conditioned, by modes of thought, to contain exact sequences of opposites. For every 'yes' contained in the upper brain system vis-à-vis a given problem, a 'no' must be equally plausible and present. In this way the binary system (the yes *or* no) may process the information of the non-binary system (the yes *and* no).

The modes of thought in these philosophies, such as Taoism, are preened to develop such sequences of opposites or contradictions, as in 'the heavy is the root of the light,' 'action is produced by non-action,' 'those who talk do not know, those who know do not talk,' and so forth. The general emphasis is on 'emptying the mind' of preconceived assumptions and sharp opinions on any subject, as in 'Is there a difference between yes and no? Is there a difference between good and evil? Must I fear what others fear? What nonsense!' The *Tao* attempts to dampen the binary system of 'yes *or* no' innate to the hormonal systemics of the upper nervous system; it is not concerned, as appears from a superficial reading and understanding of these philosophies, and as unfortunately was read into these philosophies by early German scholars of Chinese philosophy, with morality in any objective way. It does not deny morality as an objective value or morality *per se*, but like all objective values it is not a point of interest. It devolves a methodology for dealing with intrinsic proclivities of the upper brain to make definite and single-pointed objectifications of anything and everything, in the interests of integrating the upper nervous system with the lower nervous system, which acts in exactly the opposite way. This too explains its preoccupation with dampening contradictions between 'yin' and 'yang' or the female and male parts of the mind and the cosmos.

From the point of view of evolutionary development the lower nervous system is associated with the emergence of the 'male type,' and the upper nervous system with the 'female type.' Thus a dampening of the contradictions between the yin and the yang or female and male parts of the individual is a means of integrating, from both a psychological and neural biochemical point of view, the upper and lower nervous systems.

The *Tao* and other later Asian philosophies take much of their philosophical conceptions and categories from the original and major philosophical system of the Orient, the *I-Ching*, known as the *Chinese Book of Changes*. In this work both the individual and the cosmos are divided into four parts which correspond almost exactly to Freud's four categories of ego, libido, superego and consciousness. The two 'male' categories are 'heaven' and 'sun' (yang), and the two 'female' categories are 'earth' and 'moon' (yin). The primal categories, those that represent priority in terms of both evolutionary development and force are 'heaven' and 'earth' while the secondary categories in relation to the parameters of evolutionary time and force are yin and yang ('moon' and 'sun'). The analogies connected with 'heaven' in this literature suggests that it is Freud's ego; 'sun' or yang his libido; 'earth' his superego; and 'moon' or yin his consciousness. The balancing factor or contradiction of 'heaven' or ego is 'earth' or superego, while the balancing factor or contradiction of 'sun' or libido (yang) is 'moon' (yin) or consciousness. In Freud's developmentary scheme, superego appears as a consequence of the development of ego in order to balance it, while the aggression of libido can be most adequately relieved or alleviated by consciousness. In the texts of *Tao* 'heaven and earth are ruthless, they treat all things as straw dogs' and 'the space between heaven and earth is like a bellows, the shape changes but not the form, the more it moves the more it yields.'

As in all religious philosophies of the 'underdeveloped' world consciousness is the highest value, and in the texts of *Tao* it is identified with yin, 'the valley spirit,' 'the root of heaven and earth.' In Freud, as well, consciousness is also the highest psychic value or component, as superego is an inhibiting factor which if malformed in the developmentary process can lead to serious maladjustment or mental illness in the individual. Both ego and libido are necessary in Freud to normal human growth and expression, but malformation of libido in the developmentary stages of individual growth lead to most of the social maladjustments in later life that are characterized by violence and brutality, while ego, if hovering between libido and superego as it does, is burdened with malformation of either of those becomes the very instrument of individual self-destruction. Inherent in Freud's psychology is a philosophy that is medical rather than properly scientific; the components of the individual's psyche are judged by criteria that devolve around the utility of these components vis-à-vis the ultimate health of the individual, and by extension, society.

Chapter V

RELIGIOUS PHILOSOPHIES OF THE FAR EAST MODEL MODERN BIOCHEMICAL, PHYSIOLOGICAL AND PSYCHOLOGICAL PROCESSES

Taoism and other similar religious philosophies of the 'underdeveloped' world extend the criteria of health beyond the individual and society to the cosmos as a whole, but they nonetheless, from a modern perspective, imply that they belong properly in the realm of medical oriented scientific and religious philosophies rather than scientific or even purely religious philosophies *per se*. These philosophies and their practices imply a relatively clear understanding of the feedback system between thought and neural biochemistry or even organ systems. The orientation of Chinese medicine, although in its specifics it continues to remain obscure to modern scientific analysis, is no doubt based on this understanding. Much as in Freud's psychology, the powers of these religious philosophies is seen to rest in the curative measures and techniques they propose, albeit their curative measures and techniques are not limited to individual or even social psychology, but extend into all phenomenological levels, encompassing the cosmos. In particular in Chinese religious philosophies the deep psychic model of God is that of a doctor, unlike religious philosophies from regions of the 'underdeveloped' world further to the west of China, where the model of God is one of a teacher, as in Judaism; a General, as in Islam; or a priest as in Hinduism or Indian Buddhism.

The fact that the individual is taught to emulate the curative powers of an archetypal God on every phenomenological level of reality is an aspect of modes of thought and communication peculiar to civilizations in the late stages of the 3rd efflorescence, where subjectification of phenomena still plays a leading role in both written and spoken forms of communication. In the languages of these regions it is with varying degrees of difficulty that conceptualization of phenomena is limited to specifics, so that the word, written or spoken, contains depths of meaning on various levels of phenomena that are difficult to separate from each other. In

Taoism and other religious philosophies of these regions the multi-dimensional aspect of words is consciously acknowledged and is given a very high value. In fact, this is one of the philosophical points made in various ways in the texts of Taoism, that is, that specificity of 'naming' or language is detrimental to a proper understanding of *Tao*, which is universal consciousness.

Such specificity or objectification of phenomena has the psychological and medical consequences of inhibiting integration of the components of psyche, and by extension leads to disease in all aspects of the cosmos, social, political, biological, natural and religious. From a modern perspective, or from the perspective of the objectification of phenomena, it is possible to recognize the analytical significance of such a scheme, once we break down the totality of the vision presented into the categories that have meaning from the point of view of objectification, i.e., into the categories of the psychic, biological, social, political, natural and religious. Conversely, by contemplating a modern scheme such as Freud's, which is by definition limited to the phenomenological level of psychic phenomena, we can see how it can be extended in the manner of subjectification and apply it in a wholesale way to all aspects of phenomenological reality. The convergence of these two methods, the objectification of the subjective or the subjectification of the objective, allows us to bridge in an intelligible way the gap between the cultures of the 'underdeveloped' and 'developed' world. It illustrates the commonality of cultures that lie on either side of the 4th efflorescence of mankind. In this way, we may make some proper judgements and predictions as to the relative rightness of those civilizations that jumped headlong without hesitation into the 4th efflorescence of mankind and those that held back on the brink and who like the 'ancient masters' characterized in the texts of *Tao*, remain, 'watchful, like men crossing a winter stream/Alert, like men aware of danger/Courteous, like visiting guests/Yielding like ice about to melt/Simple, like uncarved blocks of wood/Hollow like caves/Opaque, like muddy pools.'

From a psychological and biochemical point of view, it is of great interest how in these religious philosophies of the late stages of the 3rd efflorescence all the phenomena of the universe is broken down with exactitude into four components. In the psychological sphere, this breakdown into four components corresponds to

Freud's analytical breakdown of the human psyche into the four tangible components of ego, libido, superego and consciousness. In the purely biochemical sphere however present technical capacities allow us to make only the rough distinction between the upper (automatic) and lower (involuntary) nervous systems, and the two neural biochemical compounds that prevail in these two distinguishable sites, and allow these systems to 'think,' i.e., norepinephrine in the upper brain and acetycholine in the lower brain. Recently, certain chemical brain compounds have been isolated, such as the dopamines, which appear to have a self-automating analgesic effect on the brain, and for this reason have been termed endomorphs ('self-morphines'). However, the general purpose of these endomorphs in systemic neural biochemistry has not yet been conclusively determined, unlike the case of acetycholine and norepinephrine which are known to act as chemical messengers in the lower and upper brain systems respectively.

The dopamines, like the acetycholines are small chemical compounds in a chain-like form, but like the norepinephrine seem to have relatively rigid spatial configurations and contain only one nitrogen atom, for one point of information, per chain. The spatial configurations of dopamine seems to suggest that when they are bound together in a long linear chain or system, as they must be in the living system, the nitrogen on one link of the chain

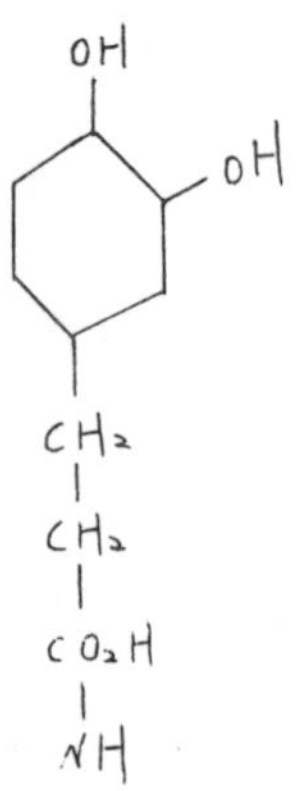

OH
OH
CH_2
CH_2
CO_2H
NH

Dopamine—Relaxed state.

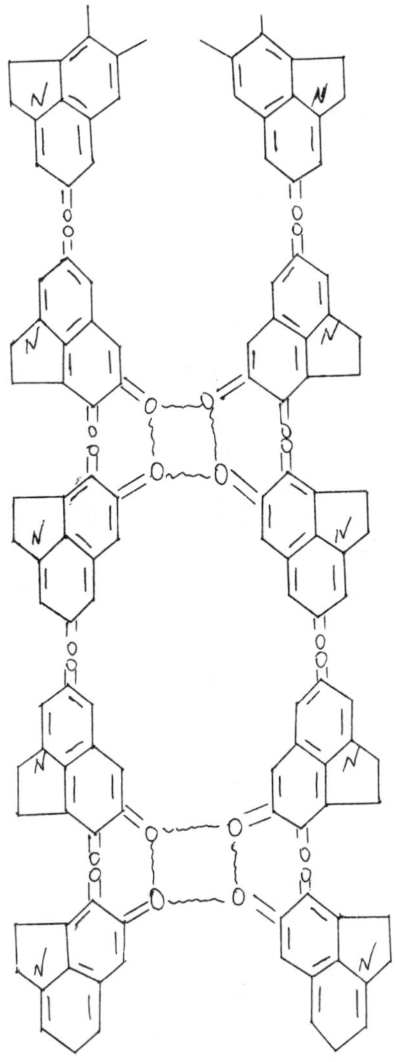

Dopamine—proposed model of contracted state. (From "Introduction to Structural Biochemistry," by Eleanor B. Morris Wu, *Chinese Culture*, Vol. XXII, June 1981.)

forms a mirror-image configuration to the one preceding it or following it. It thus seems that dopamine represents a chain of contradictory opposites. While in terms of symbolic logic the structure of acetycholine represents 'yes' *and* 'no,' and the structure of norepinephrine represents 'yes' *or* 'no,' the structure of dopamine represents 'yes' *or* 'no' *and* 'no' *or* 'yes.' While the acetycholine of the lower nervous system represents the co-existence of duality or contradiction, and the norephinephrine of the upper nervous system represents the separation or particularity of information, dopamine represents both of these simultaneously.

Duality of contradiction exists, but at the same time it may be isolated and separated, so that a consciousness of duality is achieved. Thus the dopamine, whose site has been found mainly to exist in the upper brain, may on the biochemical level be the component of consciousness. The curative value of consciousness in the texts of Taoism thus conforms to the known scientific facts of the analgesic effect of the dopamine. Norepinephrine with its site in the upper brain system and its single informational value representing a definite 'yes' *or* 'no' is the biochemical component of superego. The upper brain system and by necessity the chemical messengers which act as transmitters of information in it are from the point of view of evolutionary development associated with the 'female type.' The ancient Chinese texts which correlate superego or morality and consciousness with the female designations 'earth' and 'moon' (yin) are, from the point of view of modern science, correct.

To understand the biochemical messenger acetycholine as the biochemical component where both ego and libido are represented, we must look at the physiological nature of the organs of the lower nervous system where acetycholine is found and whose functions it directs. Found in the laboratory, acetycholine is a non-structured chain, but there is every reason to suppose that under certain conditions in the living organism this unstructured chain takes on quite a rigid formation, similar to that of norepinephrine and dopamine in the upper nervous system. The earliest forms of the lower nervous system, such as the stomach of an organism, has two basic modes: ingestion and expulsion or excretion. As organs of the lower nervous system in evolutionary time become

$$CH_3-\overset{\overset{\displaystyle O}{\|}}{C}-O-CH_2-CH_2-N^+-CH_3-CH_3-CH_3$$

Acetycholine—Relaxed state.

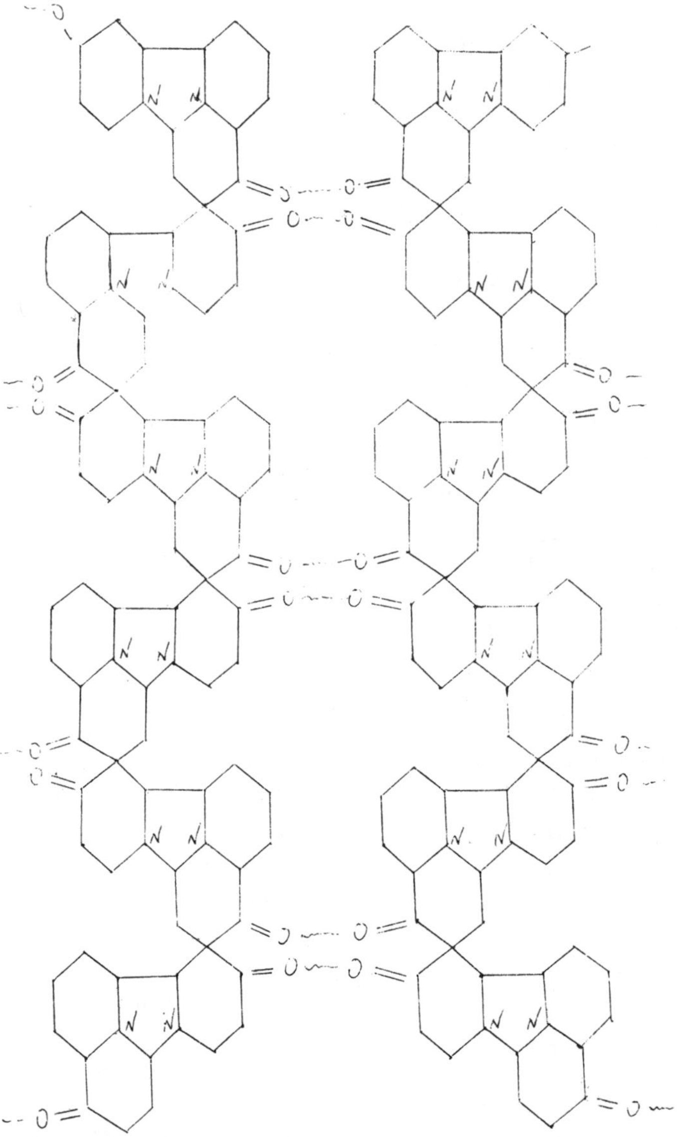

Acetycholine—Proposed model of contracted state. (From "Introduction to Structural Biochemistry," by Eleanor B. Morris Wu, *Chinese Culture*, Vol. XXII, No. 2, June 1981.

more differentiated these modes can be characterized more generally as contraction and relaxation, as in muscle systems in general.

The contraction of an organ system represents the organ in the mode of fulfilling its purpose vis-à-vis the total organism, while the relaxation of the organ system represents the path of least resistance for the organ, and in this sense represents its mode of fulfilling its purpose vis-à-vis itself. As the chemical messenger which directs the purpose of the organ vis-à-vis the entire organism, acetycholine in a rigid state carries the specific information that directs the purpose of that organ. Like the contracted organ it directs, acetycholine in a rigid chemical structure maintains an equilibrium between the integrity of the organ and the purpose of the total organism it serves. When the chemical messenger reverts to a non-rigid structure, like the relaxed (or even moribund) organ in which it is found, its purpose is no longer inter-organic; information and purpose both are diffused. While in the contracted organ, rigid acetycholine has information and purpose as its main mode, in the relaxed organ non-rigid acetycholine carries no specific message and has no specific mandate or purpose.

In its relaxed state, acetycholine possesses duality and displays co-existence of contradictions. In its rigid structure where both acetycholine and the organ it directs are contracted, specific information is conveyed. It would be too simplistic to say that in the contracted state organs and their chemical messengers are active, while in the relaxed state they are reactive, since a contracted organ may well be either an active or reactive organ *per se*. On the range between contraction and relaxation it might be better to use the terminology 'directed' versus 'primal' where when the organ system and its chemical messengers are 'directed' they have purpose vis-à-vis the total organism, and where they are 'primal' they have purpose only vis-à-vis themselves. The necessity of dual informational systems for 'directed' organs and messengers is that even when 'directed' the organ has purpose vis-à-vis itself, as well as the total organism. Thus a dual system of contradictory information is necessary for organs and chemical messengers of the lower nervous system, even at maximum purpose or 'consciousness.' This dual system of contradictions is relaxed and diffused when the organ and its messengers are in a relaxed state, as all information follows the path of least resistance, i.e., the purpose directing the organ vis-à-vis itself only and not in relation to other organs or the total organism.

The psychological parallels are clear: ego corresponds to the organ and its chemical messengers in the 'directed' mode and libido to the organ and its chemical messengers in the 'primal' mode. As characteristic of the lower nervous system, these dichotomies are associated with the 'male' type in evolutionary development and conform to the masculine designations of 'heaven' for ego and 'sun' or yang for libido in the scheme of ancient Chinese philosophy. Here chemical messengers of the lower nervous system, organ morphology of the lower nervous system, and the psychological components of ego and libido can be seen as parallel series, giving an added dimension of credence to the ancient Chinese scheme. Biochemical messengers are important since they represent the newest dimension of evolutionary theory which are particularly important in the phenomenological aspects of humanoid efflorescence.

Ego attains 'consciousness' of a sort in relation to all the other psychological components of self, i.e., superego, consciousness, libido, but the link with libido is more powerful than with other components of psyche since libido represents the essential 'purpose' of ego—the fulfillment of primal individual instincts and needs. When ego is relaxed, or non-directed or non-conscious, it is overwhelmed by libido, the path of least resistance for ego. Thus libido may be considered the 'primal' mode of 'directed' ego. In the ancient texts of the 'underdeveloped' world, ego corresponds to 'heaven,' libido to 'sun' or yang and both are given masculine gender designations, while superego as 'earth' and consciousness as 'moon' or yin are given feminine gender designations. From the point of view of evolutionary development the 'male' type precedes the 'female' type in evolutionary time, the 'female' type being a development that does not even occur in evolutionary history until mammalian efflorescence and which does not come into its own as a special distinctive type until humanoid efflorescence. All of humanoid evolution can be gauged according to the degree of development of the 'female' type in the specific efflorescence. Only at the point where the 'female' type is distinctively developed in humanoid evolution do superego and consciousness as psychic

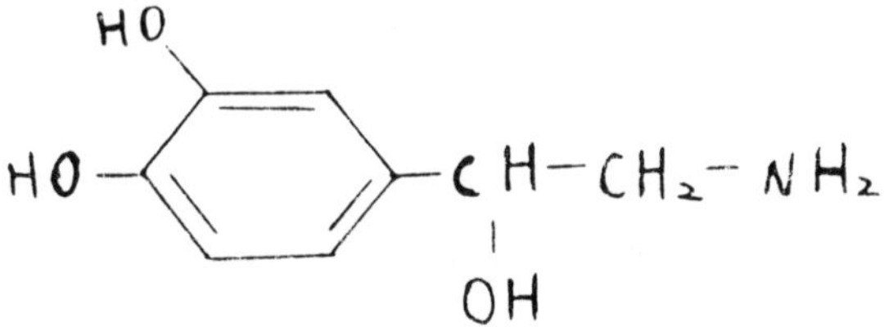

Norepinephrine—Relaxed state.

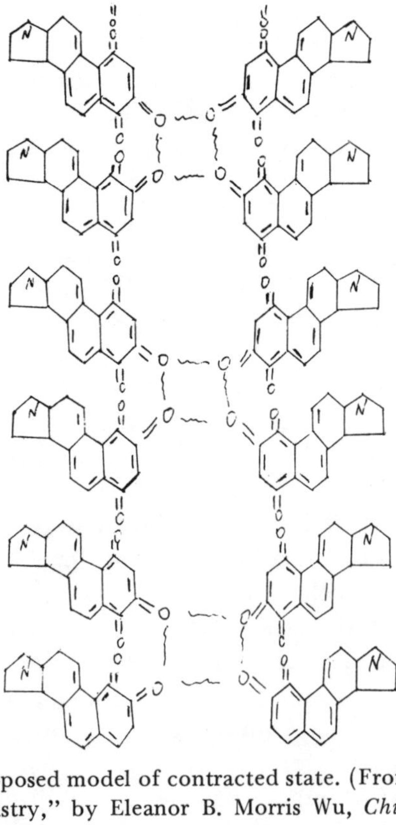

Norepinephrine—Proposed model of contracted state. (From "Introduction to Structural Biochemistry," by Eleanor B. Morris Wu, *Chinese Culture*, Vol. XXII, No. 2, June, 1981.

modes among humanoid collectives appear. The degree of development is measured by the degree to which the 'male' type absorbs aspects of the 'female' type. Ego and libido development on the other hand are developments that precede in humanoid evolutionary history the development of both superego and consciousness. They appear as developments in humanoid evolution before the 'male' type has had adequate opportunity to absorb the characteristics of the 'female' type. As a matter of fact the development of ego and libido inhibited subsequent development of superego and consciousness until modes of harnessing ego and libido were institutionalized in human collectives through systems of slavery.

Prior to the institutionalization of slavery the 'male' type with institutionalized ego and libido was the primary gender type in humanoid collectives with little or no influence from the 'female' type. Thus the classification of ego and libido as 'male' types in the ancient texts of the 'underdeveloped' countries is, from a scientific evolutionary point of view, also correct. The lineage between ego and superego, as advanced by Freud, as developing in counterbalance to each other corresponds to the linkage placed between 'heaven' (ego) and 'earth' (superego) in these ancient texts, and can also be seen to be correct from the scientific evolutionary point of view. As ego is the 'direction' of primal psychic energy, it corresponds to the 'directedness' of organs and their chemical messengers, and is thus the most 'conscious' aspect of the lower nervous system. Superego on the other hand is the least conscious aspect of the upper nervous system. While ego corresponds to the inter-organic directedness of the lower nervous system, or organ directedness vis-à-vis the organism, superego corresponds to the intra-specific directedness of the organism. The purpose of the organ in its directed state is to conform to the overall purpose of the entire organism, while the purposeness of the organism in its directed state is to conform to the overall purpose of the collective or the species.

Ego and libido are the psychic componants of the lower nervous system. In relation to the physiology of the organism, the lower nervous system represents in general the priority of the organ over the organism. The most directed or conscious aspect of the lower nervous system is the organ in a contracted state, and the biochemical messenger it contains in a rigid state. In the contracted and rigid states of the physiology and biochemistry of the lower nervous system, a kind of consciousness is achieved which has access to the purpose of the entire organism. On a psychic level, ego is able to channel primal psychic drives to generalized needs of the entire organism and thereby achieves a measure of consciousness. All aspects of the lower nervous system as opposed to the upper nervous system are distinguished by these differing contexts.

From a Freudian point of view, all aspects of the lower nervous system are in a state of narcissism. The primal biochemical messengers of the upper nervous system, norepinephrine and its psychic correlative, superego, on the other hand are entirely out of context; they are discontinuous from the purpose of the organs which are ultimately responsible for sending all information into the upper brain. Norepinephrine and superego make distinctions on a 'yes' *or* 'no' basis only, and may thereby be entirely removed from the purposes for which these decisions were ultimately made. Out of touch with organ or psychic systems, norepinephrine and superego may be as destructive to the organism as a whole as naked ego or unharnessed libido, as where superego is concerned Freud goes to great pains to demonstrate. The integrating psychic factor is consciousness, whose biochemical correlative may be the dopamines, for only through consciousness can decisions be made that are both in context of organic systems and out of context of organic systems simultaneously. The biochemical compound dopamine provides the information system by which this can be achieved, as specific isolated decisions can be made that are nonetheless linked on a chain to their contradictory opposites and necessary dualities. The contradictory opposites and necessary dualities put the out of context decisions ('yes' *or* 'no') into context ('yes' *and* 'no') but not at the same point in time and space. Context is appreciated but not at the expense of specifics. In this way both the overall health of the organ and the organism is maintained, and the overall health of the individual and the collective is maintained. Consciousness allows collective decisions to be made, but never at the expense of the collective.

In the texts of Taoism, consciousness is associated with 'moon' or yin, the reflected light of the sun in the heavens and on the earth. Through adherence to 'yin' *Tao* can ultimately be achieved, where 'when nothing is done, nothing is left undone.' In the practices developed out of religious philosophies such as Taoism in the 'underdeveloped' world, not only is the individual never sacrificed for the collective or vice versa, but the organ is also never sacrificed for the organism and vice versa. All functions of the lower nervous system are infused with consciousness, and all functions of the upper nervous system are infused with the primal psychic and physiological forces of the lower nervous system. A perfect state of physical and psychic integration is achieved, giving extraordinary powers to an individual trained in these practices which can be harnessed for the purposes of the collective.

Nonetheless, analyzing as we are a system that derives from subjectification of phenomena in terms of objectification of phenomena, certain nuances and implications of the resulting system are found to be inadequate, if not misleading. Religious philosophies of the 'underdeveloped' world do not break down phenomenological levels of reality into the psychic, physiological and biochemical componants of reality as we are mandated to do. In these philosophies and their resultant practices, these levels exist on a continuum, where there is no clear separation of one level of reality from another, and where both intuition and cognitive observation work together to enlarge the system to ever more expanding levels of complexity on the one hand and ever more precise particularities on any given level on the other. In the development of these philosophies, it goes without saying that there was no construction of a rigid system of psychology, physiology or biochemistry as such, or as we understand them today from a modern perspective. Yet, for all intents and purposes, an understanding of these various systems exist in these philosophies, however the particularities of one system blend in a continuous way with the particularities of another. In addition, insofar as these philosophies are products of civilizations in the later stages of the 3rd efflorescence and not the 4th proper, there is no well defined development of collective ego and libido as such. On the contrary, these philosophies have the teleological end of negating ego, whether individual or collective, and abdicating libido, whether individual or collective.

The psychic and physical integration which is achieved in these philosophies necessarily extends beyond both the individual and the collective self to the 'self' of the cosmos, or God or the power of creation. Since the 'self' of the cosmos or God is equated with consciousness, which takes into account the context of all levels of phenomena, the purpose of cosmic self or cosmic consciousness is beyond the interests of the individual or the humanoid collective. Negation of ego and abdication of libido are of necessity the goals of an individual in tune with cosmic consciousness. This negation of ego and abdication of libido is a goal not only of the individual self, but also of the collective one. Thus, if the physical and psychological integration of the individual that is achieved through the practices of the philosophies is to be harnessed for the purposes of the collective ego and libido, the nature of collective ego and libido, at least from the point of view of the individual who serves it, must be seen to be self-negating and self-abdicating.

The will of the collective must be seen to be as pacific as the will of the individual so trained, so that total harmony on all phenomenological levels, which allows this extraordinary integration in the first place, may be maintained. On the other hand, since ego and libido must be acknowledged before they are negated and abdicated, ego and libido are displaced into their negation and abdication reference points, i.e., on the first level, individual consciousness through ultimately to cosmic consciousness. The scale of displacement is so enormous that a power results that may be characterized as anti-ego and anti-libido. That is, the displacement of the primal psychological force of ego and libido is in the superego and the consciousness and exists there, albeit in a negative way from the point of view of ego and libido as such. Thus, while the resulting collective will is pacific it is also imbued with an absolute sense of self-righteousness and omniscience, as the collective will is identified so closely with the cosmic will which, in taking into account the context of all phenomena, is ultimately responsible for the creation and activity, the birth and death, of all things.

While the collective will of civilizations in the late stages of the 3rd efflorescence is not organized towards specific collective actions and goals as are civilizations of the 4th efflorescence, when collective will is manifested it is imbued with a finality and sense of teleology that is lacking in civilizations of the 4th efflorescence.

In a sense to organize this collective will in any way other than one that appears accidental would appear immoral from the point of view of these religious philosophies, as organized collective will would smack of a *hubris*, individual and collective, that has an extremely pejorative value in these philosophies. So-called accidental organizations of the collective will on the other hand, being beyond the control of individual cognition, are imbued with the teleology associated with the cosmic self, and would thus have the full assent of both the individual and collective self.

The ancient texts of the Chinese *I-Ching* or *Chinese Book of Changes* presents a balance between the 'male' ego and the 'female' superego in the guise of 'heaven' and 'earth.' They also present a balance between the 'male' libido and the 'female' consciousness in the guise of yang ('sun') and yin ('moon'), much of which is made of in the texts of Taoism and other religious philosophies of the 'underdeveloped' world. It is possible to analyze the balance between ego and superego in modern categories of psychology, physiology and biochemistry, but the balance between libido and consciousness stretches the parameters of these categories to extremes which are not very credible. To understand this balance from a modern point of view, it is necessary to extend our modern psychological, physiological and biochemical systems into realms of both the natural and teleological, which extension may be understood in terms of analogy if not in the objectification of phenomena *per se*. In fact we may characterize much of the thinking of these religious philosophies as reasoning by analogy, and although this reasoning by analogy is not analyzed as the methodology of analysis or synthesis in these religious systems of the 'underdeveloped' world in any self-conscious manner, such reasoning in and of itself does not stretch the boundaries of objective analysis in any objectionable way. Thus, libido, as the primal psychic energy, may be associated with the movement of the universe, and all of its activities, while consciousness on a cosmic level is associated with the mind of God that directs this movement and activity. While the consciousness of the individual takes everything into account in context, the mind of God is so vast and complex that even while it takes all movement and activity of the universe into account in context, its processes of thought are beyond the scope of the individual mind.

The belief in a cosmic consciousness allows the individual to accept all the seeming contractions and movements of the universe as sensible and rational, since that is the nature of the cosmic consciousness, even if the individual is unable to see either the sense or the reason of these movements and activities. Such a faith in the cosmic consciousness allows the individual to accept with a great deal of passivity the movements and activities of all phenomena, even when these movements and activities seem contrary to both his ego, or individual self interest, and superego, or individual sense of morality. In *Tao*, one is at one with the universe, that is, one accepts the ebb and flow of changes and contradictions in the universe without either despair or elation, secure in the sense that such changes and contradictions are caused by the super-consciousness or the cosmic mind. Applied to the level of individual existence, attainment of consciousness allows the individual to accept the primal energies of libido, which also often seem at odds with ego or self-interest, or superego, where a sense of morality is socially rather than either individually or cosmically imposed.

From a psychological point of view, this kind of natural acceptance of libido is what is so often necessary, in the treatment of people with psychological problems, for attainment of psychological balance and health. Acceptance of that which cannot be always understood in terms of ego or superego is a necessity in terms of individual balance. Componants of the libido such as aggressive drives and sexual instincts if not naturally accepted by consciousness, without the tedious pedantic apologies of the superego, will lead to serious personal and social maladjustment. On a collective level this contradiction has similar constructs, so in the texts of *Tao* it is said, 'Tao is great/Heaven is great/Earth is great/ The King is great.' The primal force of the collective, corresponding to the primal force of the individual libido is authority, and this too must be accepted by consciousness. From the point of view of modern psychological science, where collective authority is not accepted, serious social maladjustment of the individual will result.

The exquisite reasoning of these religious philosophies of the late stages of the 3rd efflorescence can not only be explained in terms of the analytical and objective categories of the 4th, but also in many cases when properly understood serve to elucidate these analytical and objective categories beyond their present level of development. It is thus entirely improper, incorrect and misguided to characterize these religious philosophies of the 'underdeveloped' world as prescientific or backward or primitive. They represent a different approach to the problem that mankind faced when he had outgrown the primary aspects of the 3rd efflorescence and was ready to move on to a future stage of human efflorescence. They represent a more hesitant approach to leaving behind the achievements of the 3rd efflorescence in favour of a new and undoubtedly dangerous venture, which would bring mankind once again face to face with the problems he had thought to have left behind in the 2nd efflorescence, albeit in a collective rather than individual manner. Thus the change from the 3rd to the 4th efflorescence is not a change from non-scientific or pre-scientific development to a scientific development. On the contrary, science was a well-established mode of behaviour in the entire continuum of humanoid development since the 2nd efflorescence. The change was far more complex and subtle, and the hesitation shown by those civilizations of the 'underdeveloped' world in racing helter skelter into the 4th efflorescence was based on the sound and self-conscious collective experience of millenia of evolution.

It was not science that was the chasm which mankind had to bridge in its transition from the 3rd to the 4th efflorescence, but the very question of human survival. The question was how mankind was to organize the forces of collective will that would work for the betterment of all individuals without at the same time unleashing the power of the individual ego and libido that would once again force all mankind to the brink of collective self-destruction. These ruminations did not emerge suddenly but were the collective teachings of all civilizations in the later stages of the 3rd efflorescence in their religious, philosophical and scientific heritage. 'To be or not to be, that is the question' as Shakespeare said: how much action would man take upon himself, how much ego-recognition, expressed through libido aggression, was possible before collective suicide would be the result.

Chapter VI

SUPEREGO AS A BALANCING FACTOR IN WESTERN CIVILIZATION

There is a general commonality of thought in all the literature of civilizations who have passed the major stage of the 3rd efflorescence which centers around the issue of 'to be or not to be,' whether or not to race into the 4th efflorescence or whether to hold back in order to develop some central strategem that might better enable humanoid collectives to meet the problems they inevitably will face in the 4th efflorescence. This commonality of thought noted by Jung in his great work on the collective unconscious tallies with modern theories of both deterministic behaviour and genetic theories. It is not that man's thought is a direct product of his genes, but if man's evolution is associated directly with his biochemical evolution, and if man's thought is at least associated with his biochemistry, then humanoids possess, at relatively similar stages of efflorescence, the potential for commonality of thought. Psychic evolution is at all points mediated by common biochemistry, and while the particulars of psychic evolution may vary somewhat from place to place and time to time, dependent as these particulars are on the specific circumstances or ecology with which a given collective is faced, the general trends and patterns of psychic evolution are more or less equivalent in equivalent stages of efflorescence. More often than not when modes of thought and behaviour vary widely from one collective to another, it is because of the variable experience of the collectives involved vis-à-vis their transition from one stage of efflorescence to the next. This is the case in point between the widely varying modes of thought and behaviour between the 'developed' and 'underdeveloped' world.

The collectives of the 'developed' world have leaped into the 4th efflorescence, while those of the 'underdeveloped' world have lagged behind in the late stages of the 3rd. The reasons for these varying modes of thought and behaviour between the 'developed' and 'underdeveloped' world are a result of the specific cumulative history and experience of these regions and collectives as they passed from the 1st to the 3rd efflorescence at different rates and under different circumstances.

The rates of transition and the circumstances which occurred from the 1st to the late stages of the 3rd or the 4th have been highly eclectic among human collectives now extant in the world. For some collectives, the transition from 1st to 3rd efflorescence happened much earlier than for others, and the duration in which they managed to survive as collectives in the 2nd efflorescence was also, correspondingly, much longer. If we suppose that all human collectives extant today entered the 1st efflorescence at about the same time, say 1-2 million years ago, as the earliest data of *Australopithecus* indicates, we may suppose that some parts of this 1st efflorescence entered the 2nd 500,000 years ago as the data suggests, or even earlier, while others did not enter this stage until 100,000 years ago at the beginning of the upper Pleistocene or even later.

From the archaeological evidence that has been found to date, we can reasonably and conservatively place a 500,000 year gap between entrance of some collectives into the 2nd efflorescence and others. The probable time disparity at which groups entered into the 2nd efflorescence and even stayed in it can be adduced from the fact that collectives of the 2nd efflorescence were able to evolve into the 3rd by virtue of institutionalized enslavement of one group by another. In all likelihood, these groups who were enslaved were always in an earlier stage of efflorescence than the enslaving group, or at least in an earlier state of the same efflorescence. Slavery during earlier periods of humanoid evolution, during the 2nd and the major part of the 3rd efflorescence was no doubt of a total nature, where the enslaved groups were no more than beasts of burden, herded like cattle, and hardly recognized as humanoid by the enslaving group. Nonetheless slavery remained as an integral part of humanoid collectives through the major part of the 3rd efflorescence, surviving even into the 4th.

In these later stages of humanoid evolution, slavery was not of a total nature, but of a political, economic and cultural one, the clearest example in historical memory being the wide-spread system of colonialism, which was a prime facet of both ancient Greek and Roman democratic imperialism and was continued into modern times by European civilizations in the 4th efflorescence. The kind of slavery instituted and carried on by the ancient Greeks and Romans was of a relatively democratic and humane nature, certainly compared to the kind of slavery carried on in neolithic times, and even by comparison with the kind of slavery well-documented to have existed in Biblical times among civilizations in the Middle East, North Africa and Southwest Asia. The kind of slavery begun by the ancient Greeks and institutionalized under the ancient Romans and modernized further by the Europeans and Japanese in recent times was of a 'tutorial' nature. That is, the enslaved group was given a period of apprenticeship, as it were, in which it was allowed to assimilate the cultural and behavioural modes of the enslaving group, at which point the assimilation of these modes being completed, the enslaved group was allowed its freedom, in the form of more or less equal status with the enslaving group.

During the period of apprenticeship however, the enslaved group was more or less at the whim, in all matters political, economic and cultural, of the enslaving group. Modernized by Europeans and Japanese in more recent times, this system of tutorial slavery reached the point of a fine art in the guise of colonialism, where the collective modes of the 4th efflorescence were spread in varying degrees to remote corners of the globe, some of which remained still in the major part of the 3rd efflorescence. The world as we know it today has been in large part shaped by the colonialism of present day Europeans and Japanese, although those civilizations in the late stages of the 3rd efflorescence have resisted with some success the assimilation of collective modes of the 4th efflorescence exposed to them through colonialism, managing at the same time to maintain equal status with the colonizing powers. The ability of such civilizations of the 'underdeveloped' world to resist assimilation of collective modes from the 4th efflorescence is not, as we have seen, by virtue of ignorance or backwardness but by virtue of genuine philosophical disagreement about the efficacy of collective modes of the 4th efflorescence as they are now practiced to ensure humanoid survival.

Slavery has never been solely an institution that was extra-territorial; on the contrary, it always existed in the first instance as an intra-territorial phenomena. Thus, in Biblical times when a group was enslaved, it would be carried off holus bolus from its own territory to serve the enslaving population in its own land. This practice was still maintained in early Greek history, but by the time of the Romans it was frowned upon, and discontinued, as it interfered with the consolidation of collective modes characteristic of the 4th efflorescence. In modes of the 4th efflorescence, social organization devolved around the development of collective ego-recognition and libido aggression, and the presence of an enslaved group inside the territory of the enslaving group presented a contradiction with which no collective of the 4th efflorescence could adequately cope. Collective ego recognition depended on the relative equality of all individuals in the collective, and if collective libido aggression were to be unleashed upon the group itself the entire configuration of behavioural modes of the 4th efflorescence would collapse. Slavery outside the territory of the collective, on the other hand, was not only permissible but also necessary if the collective were to express collective libido aggression.

In these early examples of the 4th efflorescence, slavery, as an individual rather than a collective mode of behaviour inside the territory of the collective, was not only permissible but also constructive in terms of maintaining modes of the 4th efflorescence. Slavery on an individual basis, where enslaved individuals were drawn from an extra-territorial colony, or from a minority group within the collective who were visibly unequal in terms of ego recognition and libido expression, such as the poor, weak and underprivileged, or female, provided a means of siphoning off libido aggression on an individual basis which might otherwise become a collectively destructive force.

During the Dark and Middle Ages in Europe and Japan, intra-territorial slavery was manifested in terms of feudalism, where subordination of those engaged in the production of goods and services, such as farmers and artisans, to a warrior elite was justified in terms of complicated religious systems that masked the realities of social, cultural and political enslavement. As these systems became nationalized to their present forms during the commercial and industrial revolutions in the 17th-19th centuries, collective ego recognition and libido aggression was no longer reposited in an aggressive warrior class that was intra-territorially operative.

Common modes of collective behaviour had allowed within the ancient Greek empire the possession of wealth and the means of production to symbolize ego recognition that had previously been only associated with prowess in war and combat and other modes of aggression. The transfer from primal modes of individual aggression to an economic representation of this aggression allowed the formation of relatively stable collectives in the 4th efflorescence, although collective ego recognition and libido aggression continued to be expressed in the constant warfare between one city state and another and institutionalized colonialism or 'tutorial' slavery practiced against non-indigenous states and groups.

Another form of colonialism, practiced still in modern times, was the Greek system of dealing with an overflow of population in a given collective or city state. Where the population became so big that the economic resources could no longer allow for a more or less equal distribution of wealth and goods among the individual members of the collective, a segment of the population was sent to settle virgin territory, or to occupy land held by a visibly weaker collective that was already enslaved to the parent group. This new settlement also was bound to the parent group as a colony, in the form of 'tutorial' slavery, where for a finite period of time, until it had assimilated the collective modes of the parent group and could act independently of it, was forced to rescind a large part of its revenue to the parent colony and to submit to it in all matters, political, social, and cultural.

As these Greek collectives were rarely able to extend their sense of collective ego recognition and libido aggression beyond the boundaries of a relatively small city state, to include all the collectives of common culture within the boundaries of a larger model, the Greek empire was plagued by constant warfare among the various city states that prevented it from extending its rationalized system beyond its borders. It might be said that the influence of the 3rd efflorescence on Greek society was still so strong that a complete development of modes of the 4th efflorescence was not possible. There was sufficient hesitancy about the efficacy of developing naked forms of collective ego recognition and libido aggression to prevent the development of a homogeneous and efficient pan-Hellenic state. While the influence of the 3rd efflorescence on the Greeks was from the major and not the late stages of that efflorescence, the Greeks were in constant contact, and indeed had derived their racial and cultural origins at least in part from, the civilizations of the Near East and the Far East where the late stages of the 3rd efflorescence were the dominant forms of collective behaviour. Collective superego as it existed among the ancient Greeks was intermediate between the cosmic sense of superego found in Oriental civilizations and the individual sense of superego that was a vestige from earlier forms of the 3rd efflorescence.

Religious worship and moral imperatives were closely associated with the tribal origins of the race and the nature of this manifestation of collective superego ultimately limited the size of the collective that could be formed with modes of the 4th efflorescence to small, city-states. Nonetheless, the exposure to the cosmic notion of superego from civilizations farther East, combined with modes of collective ego recognition and libido aggression that resulted from the historical experience of the Greek people, produced the notion of a pan-Hellenic empire, where modes of the 4th efflorescence were seen to be applicable on a universal basis, and which modes had strong moralistic overtones. These strong moralistic overtones were the result of a synthesis of total Greek experience of both the major and later forms of the 3rd efflorescence, combined with the 4th. Collective ego recognition and libido aggression was never viewed as an end in itself, existing only for the welfare of the collective involved, but was imbued with a strong sense of collective superego and consciousness and embedded firmly in individual superego and consciousness associated with the tribal origins of this people.

As a result, collective ego recognition and libido aggression was always rationalized in terms of the philosophical notion of justice, and it was this notion that presented the contradictions which naked collective ego and libido could not overcome and which ultimately led to the permanent fragmentation of the Greek empire.

The foundations of Roman civilization were based in the colonies founded by the Greeks in southern Sicily probably thousands of years before the rise of Rome. Thucydides tells us that the Athenian invasion of Sicily (about 500 B.C.) failed, where Athenian invasions on Greek territory had succeeded, because the Italian society that had arisen from the melange of former Greek colonists and indigenous peoples who came from the North, the East and the South, from Europe, Asia and Africa, were far more able to cooperate among themselves and accept centralized authority and direction than were the Greeks themselves. Over the period of time during which Greek colonists had settled on the Italian peninsula, separation from their home territories with the mythical and magical associations that superego ascribed to specific territorial locations had dimmed the links of these people with the major part of the 3rd efflorescence. Furthermore, separated by formidable geographical barriers from the civilizations to the East, these new Greek settlements were no longer in constant interaction with people who held the notions of cosmic superego and cosmic consciousness.

Existing among the melange of indigenous people who inhabited the Italian peninsula, as colonists they had the time and the superior modes of collective behaviour that allowed them to inculcate these peoples in the forms they had brought with them from Greece, and to refine these forms, unhindered and uninhibited by conflicting and contradictory modes of both individual and collective superego. The notion of justice associated with collective ego recognition and libido aggression was rarified and attained the status of a philosophical system which was manifested in practice through inculcated and refined modes of behaviour, on both individual and collective levels. As Roman civilization extended northwards and westwards, it extended these practices among those whom were included in the Roman empire, and 'tutorial' slavery lost much of its pejorative aura, the gap between the enslaved and the enslavers being little more than academic. Enslaved groups under a period of 'tutorialship' were encouraged to discard superego modes of the 3rd efflorescence, and to bind themselves in a universal, rationalized culture with the Roman colonizers. The efficacy of Roman extension is a rather clear matter of historical record. Where the historical experience of colonized collectives did not differ too markedly from those of the colonizers, inclusion in the Roman Empire and the success of the *Pax Romana* was ensured. However at the point where Roman civilization encountered the Goths and Visigoths of the North East and North West of Europe, not only did the *Pax Romana* fail, but the entire Roman system was placed in jeopardy, and ultimately, under pressure from encounters with these peoples, collapsed.

The central Asian, or Germanic, hordes derived from humanoid collectives that were nurtured in the harsh and inhospitable climes of northern Europe and northern Eurasia. From both historial documentation about the collective modes of these peoples, and from the extant Norse myths which comprised their philosophical and religious notions, it is clear that the transition from the 2nd to the 3rd efflorescence among these peoples was erratic. The harsh climates and difficult ecological niches where they evolved made modes of the 2nd efflorescence adaptively advantageous where in warmer and more hospitable climes and niches such modes proved clearly to be the obverse. Where sources of food, shelter and warmth were hard won in the northern climes, attributes of libido aggression helped preserve the sparse and loosely knit collectives of these regions, for such were the environmental circumstances in which they lived that libido aggression could be largely unleashed against natural circumstances for the purposes of sheer physical survival, rather than against other humanoids, at least to the extent that self-extinction of humanoid collectives did not become a distinct danger. In fact, without powerful energies of individual ego-recognition and libido aggression, survival of collectives in such harsh natural circumstances would have been an impossibility.

Thus, even when these collectives in time entered into the third efflorescence, modes of the 2nd efflorescence remained cherished values, signifying as they did the moral imperative of group survival. In the Norse myths, female cult figures and personages such as Signy, Brynhild, Frigga and the Valkyries were cold, heartless, and secretive, and as prone to battle and as able as their men folk. There is little evidence of a well-developed maternal sense in the female cult figures of these peoples. Out of allegiance to her father's lineage, Signy kills her husband and her own children and then herself. Similarly Brynhild, out of revenge, kills both her lover and her husband and herself, also in a remorseless way, while Frigga, the chief Goddess of the Norsemen, was secretive and covetous. Freya, the Goddess of love and beauty, claimed half the dead of any battle for her own, and the Kingdom of Death belonged to a female personage, Hela. The male personages and cult heroes did not so much love and adore their female counterparts, as they were jealous of them for their equal prowess in war and their secret powers. In short, the Germanic hordes maintained a belief system where the 'female' type was ill-developed and barely differentiated from the 'male' type in keeping with modes characteristic of the 2nd rather than the 3rd efflorescence.

Although institutionalization of superego existed among these groups, as can be adduced from the multitudinous facets of civilization they possessed common to the 3rd efflorescence, such as language, written and spoken, rudimentary medicine, pharmacy, religion, cooking, etc., superego never acquired the stature it attained among other collectives in more hospitable climes and ecologies. The 'male type' remained the most valued of the sexual types, and this value was assented to by both males and females of the collective alike, since male aggression did not imperil the security or survival of the females or the offspring of the collective, and was in fact a necessary adaptation for the collective as a whole. As these races acquired the accoutrements of the 3rd efflorescence and some concept of collective will, they expanded their search for the material and environmental circumstances that were advantageous to the group as a whole in more southern and warmer climes. The method of acquiring better material and environmental circumstances was primarily one of aggression against both environment and the more settled and less aggressive groups who inhabited these warmer climates. Aggression as a method of acquiring better material and environmental circumstances became institutionalized, as environment, circumstance and experience gave them little opportunity to develop the settled agricultural modes of collective behaviour common to other groups who had evolved from the 2nd to the 3rd efflorescence under more favourable circumstances.

Thus, these Germanic hordes acquired many of the modes of the 4th efflorescence without having been firmly established in the 3rd, or ever having been exposed to modes common to late stages of the 3rd efflorescence. The combination of highly valued vestiges of the 2nd efflorescence in tandem with newly acquired modes of the 4th made them a fearsome historical force. Roman civilization, whose efficacy in colonizing was dependent on the colonized groups having both a well-developed superego that could be tutored into a universal sense of justice and collective cooperation as well as less well-developed modes of collective ego than themselves, was helpless against such a force. In addition, as Roman civilization extended itself both in Europe and in Asia, it was exposed more and more from the Asian side to the cosmic moral imperatives which weakened the aggressiveness of their collective will.

The overwhelming aggression of the Germanic hordes increasingly brought Roman civilization face to face with its own collective aggression, which in the end it could not tally with its own notion of universal justice, for as this notion became increasingly separated from territoriality and the individual superego associated with such territoriality, it became rarified to the point where it blended with the notion of cosmic consciousness and superego advancing towards Roman civilization from the East. Since the objectification of phenomena was established in Roman civilization and its benefits were clearly recognized in the value of material welfare it offered to the mass of its citizens, collective modes of the 4th efflorescence were never abandoned in favor of the late stages of the 3rd efflorescence. While there was no turning backwards for Roman civilization, there was no going forwards either, and Roman civilization stagnated in the face of these insurmountable contradictions. In the Roman way a compromise was achieved between the opposing forces of the Germanic reality, their own self-doubts, the visible accomplishments of their own civilization and the undoubted moral superiority of Eastern philosophy. Christianity was first accepted and then developed by Roman civilization as a method of colonizing the Germanic invaders by spiritual means, where the means of collective aggression had failed.

Militarily the Germanic hordes had remained undaunted, as their poorly developed institutionalized superego had made them unafraid of defeat or death, and no matter how many times they were vanquished were ready to fight again regardless of the cost to the collective, since the very sense of the collective resided in an individual and collective ego-recognition whose only major criteria was libido aggression and was devoid of a sense of the consequences for such aggression in terms of group stability and security. The Germanic hordes were equally impervious to the economic benefits of peace and collective cooperation, since those modes of the 4th efflorescence which allowed economic power to be substituted for individual and collective aggressiveness had never developed. In the end, for the price of conversion to Christianity, the Romans surrendered the material and territorial wealth of their civilization to the Germanic hordes, and the lengthy and painful process of evolving European civilization began.

European civilization, as conceived by the Romans, kept the collective modes of the 4th efflorescence as it had been developed by them in ancient times under the adjudicating influence of the morally superior religious and philosophical system of Christianity. Just how lengthy and erratic the process of Europeanizing Nordic and Central Asian hordes has been is a matter of record in our own times. Under the philosophy and reign of the Nazis in Germany during this century a revival of primordial Nordic culture arose that threatened once again to smother the civilizing efforts of Roman and Christian civilization. By the 20th century, the Germanic peoples, having attained the status of nationhood, seemed to be the recipients of enormous benefits brought to them by Roman and Christian civilization. Collective modes of the 4th efflorescence under the adjudicating moral wisdom of Christianity had allowed Germany to become the very model of a modern European state, where with the ascendancy of moral sanctions every aspect of developed civilization was refined and flourished, in the arts, sciences, technology, medicine, industry, law and commerce. The greatest theoretical minds of the 4th efflorescence since the time of the ancient Greeks were produced in this environment, such as Freud, Jung, Einstein, Marx, Heisenberg, Planck just to name a few. Nonetheless, underneath the surface of the society neither the collective modes of the 4th efflorescence as developed by the Romans nor the moralistic adjudicating influence of Roman Christianity had been well established among the individuals of the society as collective modes of thought and behaviour.

All of the refined arts, sciences and technologies of a developed society were self-consciously turned and used to promulgate the modes of the earlier, primordial Germanic hordes. Objectification of phenomena in the sciences, art, and technologies was used to justify in a causal way these modes. Ego-recognition, or libido aggression on both individual and collective levels was promulgated with the exquisite logic adopted from Grecian dialectics, and the cold, heartless, unmerciful and aggressive 'male type' was mythicized in terms of expository logic. The kind of slavery which had been unknown in humanoid collectives since the middle Pleistocene was institutionalized, and racially identifiable groups such as Jews, certain kinds of Slavs, and engaged Christians were singled out as 'sub-human' on the basis of their pacific proclivities, to be used as beasts of burden or exterminated *en masse* at the whim of the 'master' Germanic collective.

Thus, Naziism glorified the primal archetype of humanoid collective which fused modes of the 2nd efflorescence with those of the 4th, having passed through the 3rd in an incomplete fashion. Nazi society seemed ultra modern because of the accentuation of the collective modes of ego-recognition and libido aggression by powerfully remaining vestiges of individual ego-recognition and libido aggression untamed by a well-developed institutionalized superego. Those groups singled out by the Naziis as 'sub-human' on the other hand represented primordial archetypes of the 3rd efflorescence hovering on the 4th, resembling those civilizations of the 'underdeveloped' world who have decided to remain in the late stages of the 3rd efflorescence for reasons that appear quite logical when we look at the destructive consequences of maladaption in the 4th efflorescence, such as Nazi society.

Many aspects of Naziism were quite consciously inspired by the example set by Russia under Stalinism. Marxism attempted to overcome the survival dilemma of 4th efflorescence modes of behaviour by postulating a society that was not merely classless, but also leaderless. The adoption of Marxism in semi-Asiatic Russia, which even in the 20th century comprised a variety of archetypes that ranged from established modes of the major period of the 3rd efflorescence, to late stages of the 3rd efflorescence, and very early forms of the 4th efflorescence, had also, due to the Nordic origins of a certain part of its most influential and 'Westernized' population, some vestiges of archetypes common to the Germanic horde type. Because of the latter, the attempt to establish both a classless and leaderless society resulted in the emergence of contradictions for which no easy resolution was possible. Notions of classlessness tallied with those vestiges of Russian society that had emerged in a continuous line from 3rd to 4th efflorescence, while notions of a leaderless society tallied with an entirely different group within the melange of Russian society, those Asiatic-type collectives that were established in the late stages of the 3rd efflorescence. Neither of these notions tallied with the modes of certain of the influential Westernized elite whose modes resembled the primordial Germanic archetype, being extremely aggressive on both an individual and collective level. The kind of society that resulted was monolithic, with a single leader, Stalin, who embodied in his leadership attributes of all these vestiges.

Those elements of Russian society who had no experience of the 4th efflorescence regarded him as a God-like figure, symbolizing the cosmic consciousness in an Oriental fashion, and whose machinations, however terrible, coincided with their conception of cosmic libido, the actions of God which were to be accepted, never questioned, and certainly never challenged. For those elements who resembled the Germanic horde model, Stalin's aggressiveness on an individual level was acceptable, and organization of the collective will on the model of such aggressiveness was considered mandatory. For those elements that had made the transition from 3rd to 4th efflorescence in a continuous and well-developed fashion, individual and collective aggression under Stalin was justified in terms of Marx's theories, where it was postulated that a certain 'waiting period' was necessary before classless, leaderless society could be established, and during which period conflict and struggle on both an individual and collective level was a necessary evil.

Nonetheless, the mode of collective aggression modeled on individual aggression which was the major reality of Stalinism was picked up by the Germans as the epitome of modern Naziism or 'national socialism.' This model has since the Naziis been used by national groups around the world and to a large extent continues to explain the kind of social and political chaos which comprises the history of our present world. There has been little understanding of the social dynamism of modern European colonialism, and for individuals and groups who harbor a variety of ambiguous archetypes, the colonialism, particularly of England and France, was mistaken for 'nationalism.' 'Nationalism' was conceived as a group well-established in a 3rd efflorescence, with its own native culture, customs and institutionalized superego, making a direct transfer into the 4th efflorescence. It was mistakenly believed that when this transfer had proven weak or incomplete that it was due to the fact that the collective involved had not sufficiently promulgated modes of the 3rd efflorescence, i.e., native culture, which modes would inevitably lead to direct transference into the 4th efflorescence, or well-established modes of collective will.

In the 19th and first half of the 20th century, this misconception was spurred by particular individuals who themselves formed an elite international clique, partaking, on the fringes, of French and British education and culture and often married into or related to ruling elites and royal families in other societies. The success of British and French colonialism was not viewed in terms of the Roman Christian model of colonization, but was seen in terms of the superior aggressive modes of British and French 'nationalism.' The sense of individual competition and aggression which arose from these misconceptions among a small group of elite power brokers developed into the 1st World War and contributed at least in part to the causes for the 2nd World War. Thus, as nationalism under Stalin, however it was a melange of contradictory archetypes and justified in terms of Marxist philosophy, it was spurred by just such an elite group who simply changed their status from elite Mandarins loyal to the Russian royal family to elite Communist party members loyal to the Soviet state. In Nazi Germany, there was so little comprehension of the Roman Christian model upon which the much envied success of French and British nationalism was based, that unlike in Russia there was no effort at all to replace abandoned Christian beliefs with any notions of universal justice whatsoever.

The disastrous social consequences of national socialism in both Germany and Russia notwithstanding, the Russians under the continuing influence of Stalinism after World War II continued to promulgate nationalism and national socialism in particular around the world, justifying such promulgation as the first stage in the Communization of the world. From the point of view of realpolitik, or collective ego recognition and libido aggression, it can be seen that Russian society, under the continued influence of Stalinism, continues to be locked in a battle to the death with foes whose shape it has inaccurately perceived. Insofar as Russia continues to believe that, as a nation-state, she is locked in battle with the nationalistic states of Britain, France, and now the United States, she not only loses credibility in terms of the philosophical system she pretends to practice, but she has also seriously misconstrued the actualities of humanoid evolution and its series of particular destinies.

Far from being 'nationalistic' movements, French and British civilizations were based on the notions of extended colonization modeled on Roman and Christian civilization. This model depended deeply on the almost scientific exactitude of Christianity to assist both their own newborn generations and the people they colonized in attaining modes of the 4th efflorescence that would not unleash undue amounts of either individual or collective ego-recognition and libido aggression. French and British civilizations were attached to the territorially associated superego in the same way as the Roman civilization they so closely modeled—i.e., only marginally. As well, Christianity assisted British and French civilization, not only to universalize both collective superego and collective ego and libido, but also to strengthen the development of the 'female type' and thereby overcome and transcend those vestiges of the 2nd efflorescence that remained entrenched in their own societies or were inherited as part of the legacy of the Germanic hordes who were encompassed in their sphere of influence.

PART II

"AN ACCIDENT OF HISTORY: THE ONTOLOGY OF HISTORICAL CIVILIZATION IN THE FOUNDATIONS OF THE EGYPTIAN EMPIRE"

Chapter VII

MANAGEMENT AND MISMANAGEMENT OF PSYCHIC DRIVES IN HUMAN COLLECTIVES

Christianity, as propounded in the New Testament and developed under the aegis of the Roman Catholic Church, treaded a fine line between collective modes of the 4th efflorescence and those of the late stages of the 3rd. The model of Christ, which Christians were taught to emulate, represented the veneration of the 'female type.' Gentle, submissive, charitable, caring and even maternal, Jesus Christ represented both a universalized superego and cosmic consciousness. But, stopping short of recognizing political authority as the manifestation of cosmic libido, Christ remained on this side of the 4th efflorescence. His ultimate crucifixion and martyrdom at the hands of political authority demonstrated a fine synthesis of rational and cosmic consciousness. By refusing to accept either the power or the authority of the state as the final arbiter of man's destiny, the example of Christ propounded a kind of rational consciousness that was independent of the phenomenological manifestations of the universe. This was cosmic consciousness expressed in the form of the objectification of phenomena, removed from subjectification of phenomena common to the concept of cosmic consciousness characteristic of modes of thought of the late stages of the 3rd efflorescence. In its objectification phase, cosmic consciousness was able to pass judgements on all workings of the universe, including political power and authority. The emulation of the 'female type' in this sense was also not merely a passive or natural judgement, whereas in Oriental philosophy the cosmic consciousness is viewed as essentially feminine, a manifestation of collective superego that gives all natural phenomena, however disastrous, a beneficial teleological end. It was, on the contrary, self-conscious and objective, a deliberate attempt to assist humanoid development with survival strategems by offering this evolutionarily necessary type as a social and psychological model.

Christianity implied that man could shape his own destiny, a destiny that was as often at odds with the workings of nature as in concert with it. As such, Christianity viewed nature and libido as forces to be managed by human beings, towards the ends of their own individual and collective survival. It was such a view that was so misunderstood by forerunners of Nazi philosophy in Germany such as Nietzche, who decried Christ as a weak and feminine creature, unworthy of masculine veneration. The management of libido as epitomized in the Christ figure was totally at odds with individuals and groups who had evolved from circumstances where individual ego and libido were closely identified with collective will, with those groups who had emerged rapidly into the 4th efflorescence by virtue of maintaining powerful vestiges of the 2nd.

By enlarging the moral base of superego and at the same time enhancing rational consciousness, Christianity was able to refine modes of the 4th efflorescence established by the early Greeks and Romans in Western civilization. An objectified rational consciousness, which had at the same time powerful moralistic overtones, was able to work on behalf of the collective ego-recognition and libido aggression while dampening their more destructive attributes. The particular family setting of Christian imagery as propounded in the New Testament and developed by the Fathers of the Roman Catholic church was particularly effective in bringing groups with some remaining vestiges of the 2nd efflorescence into more stabilized modes of the 4th efflorescence. In his family setting, with a venerated mother and a natural father who, in playing no actual role in Christ's conception, obviated much of his ego-recognition and libido aggressive affect in the family setting, combined in the context with loving brother-disciples, the model of Christ provided an effective means for superego to be broadened and more firmly institutionalized, allowing it ultimately to take on cosmic and universal associations.

In such groups where collective superego was weak and represented early phases of the 3rd efflorescence, rather than late or middle phases, and where the 'female type' was poorly developed, the imagery of Christ which associated him with the roles or in the context of doctor, priest, philosopher, fisherman, farmer, cook and sorcerer, allowed a fuller development of modes of the 3rd efflorescence, upon which basis later stages of the 3rd and the 4th efflorescence could be developed. Christianity, of course, has proven much less effective among groups in the middle or late stages of the 3rd efflorescence where vestiges of the 2nd efflorescence are few. In these groups, superego is already well-established and even broad, so that the powerful psychological effect of the Christian imagery, oriented towards libido mitigation, has little or no significance.

However, from the point of view of evolutionary development, emergence of Protestantism in Northern Europe starting in the 16th century, where collectives derived from the original Germanic hordes were established, represented a regression away from the civilizing influences of Roman civilization. Protestantism stripped away the class of celibate Priests who by the very example of their lives and work represented the 'female type' in masculine guise, encouraging the 'male type' to forego at least to some extent the more destructive modes of ego-recognition and libido aggression in his individual and social behaviour. This class of celibate priests had embodied a universalized superego which at the same time, in practical terms, acted as surrogate maternalistic figures to all members of society, alleviating thereby the lack of maternal parenting as accident and inherited regressive modes of the 2nd efflorescence wherever they occurred. Elimination of the Roman church bureaucratic system not only deprived these Protestant collectives of the surrogate mother embodied in the confessor and parish priest, but also the surrogate of the benevolent father figure embodied in more aloof forms of the clergy, the bishops, archbishops, and the Pope himself. The doctrine of the Elect had much the same psychological impact as the ancient Nordic myths of Sigurd and Brynhild: the individual was isolated from divine intervention through any human agent, and was left in perpetual doubt as to his ultimate teleological end.

As the individual and by extension the collective, could only hope for salvation through action and works, individual ego-recognition and libido aggression became once more of primary importance in Nordic society. The moralistic imperatives attached to action and works, or ego recognition and libido aggression, were slim, and entirely at the discretion of the individual superego, rather than any institutionalized embodiment of the superego. As Max Weber noted in his famous work, *The Rise of Protestantism and the Birth of Capitalism*, Protestantism unleashed an enormous amount of individual ego recognition and libido aggression in countries of Northern Europe in morally neutral economic modes of behaviour, and was closely identified with the phenomenally rapid industrialization in Northern Europe, while in more southernly, Catholic Europe industrialization was by comparison much slower.

Individual economic success and its cumulative effects was one of the forces in Germany that reinforced collective modes of the 4th efflorescence in tandem with powerful vestiges of the 2nd and which inspired centralization of an aggressive German state under the iron rule of Bismark in the 19th century. The unbridled economic activity among Germans caused by the increase of individual ego-recognition and libido aggression unleashed by Protestantism emerged hand in hand with a corresponding scientific and technological development. In Roman Christianity, science, like all accoutrements of civilization, was never viewed as an end in and of itself. Instead, it was seen from the perspective of rational consciousness which viewed human progress and collective stability as the final end of civilization, towards which all of the accoutrements of civilization were to be formulated. If they did not meet both this requirement and the requirement of universalized superego and morality as an end in itself, they were prohibited, such prohibition often being decried by modern thinkers as the reason for what they view as the undue lag in scientific development in Western civilization during the 16th century when Roman Catholicism was still in ascendancy in Europe.

With the rise of Protestantism in Northern Europe, science was no longer viewed in the framework of a rational consciousness which judged it in terms of its ultimate utility for the benefit of stable collective modes, or in terms of universalized superego which judged it according to moral worth. Instead it began to be viewed simply in terms of the increased power it yielded for the collective, thereby again reinforcing vestiges of the 2nd efflorescence superimposed on modes of the 4th. Science was viewed in terms of the increased ego-recognition and libido aggression yielded for the collective, and individual success in scientific and technological invention and development was rewarded by the collective in increased ego-recognition and opportunity for libido aggression for the successful individual through increased social, cultural, economic and political rewards. In time, the pursuit of science for its own sake so well rewarded by increased ego-recognition and opportunity for libido aggression for individuals engaged in this pursuit, became extended to fields outside of natural science *per se*, to fields that had formerly lain entirely within the province of religion, philosophy, and art, provinces formerly belonging to the sacrosanct universalized superego.

On the positive side, science for its own sake gave rise to a fresh rationality which when applied to the dominions formerly belonging to the universalized superego, resulted in new fields of study such as psychology, sociology, anthropology, archaeology, and historeography. When tempered by strong superego in individual cases such a fresh rational approach to realms once sacrosanct produced the greatest scientific and philosophical discoveries and developments in Western civilization since the time of the ancient Greeks. The works of Freud, Jung, Marx, Simmel, Weber, Durkheim, to name just a few, are among these, where fields of study that had previously remained in a realm of relative subjectification were clarified in terms of the objectification of phenomena. On the negative side, where individuals involved were not endowed with strong superegos or an independent rational consciousness, science for its own sake or objectification of phenomena for its own sake eliminated rational consciousness and morality from a wealth of social and cultural products which had been the legacy of Roman, Christian civilization. Nietzche was the unwitting leader in this direction and the trend was carried to its logical extreme by Stalin, Rosenberg, and Hitler, and virulent forms of nationalism and national socialism were the result of this process in the social, economic, political and cultural provinces of civilization, while in the scientific and technological provinces development of nuclear military capability was the ultimate result.

Actually, science for its own sake or objectification for its own sake was the path of least resistance for individuals and collectives with entrenched vestiges of the 2nd efflorescence superimposed on nebulously formed modes of the 4th. The problems of weak collective superego could be bypassed or dispensed with altogether, and problems of weak individual superego could be justified in terms of objectified individual and collective ego-recognition and libido aggression. Nationalism as an ideology objectified collective will without any regard to universalized superego or rational consciousness. More often than not, nationalism as an ideology has devolved into national socialism, where superego such as it exists is associated with the territorial and tribal attributes of the 'nation' involved, thereby justifying the absence of any universalized superego. In these ideologies again the path of least resistance is taken, since with all extant collectives in the world some phase of the 3rd efflorescence, however sketchy, has been passed through and exists as part of the collective consciousness. Even in societies with powerful vestiges of the 2nd efflorescence intact this is the case, and is part of the basis on which modes of the 3rd efflorescence have either been built, or can, theoretically, be developed.

The slim moralistic attribute of societies that have not fully passed through the 3rd efflorescence can thus be devolved entirely upon the modes of the collective itself, and do not demand any moralistic sensibility towards individuals or groups outside of the particular collective involved. Such is the morphology of national socialism itself, where collective and individual ego and libido are mutually reinforcing and are mediated by superego limited to the parameters of the collective. A collective superego limited to the parameters of the collective in turn helps in bolstering individual superego, which in such a society, is otherwise weak, and which without such bolstering from a limited collective superego threatens the very survival of the collective through destructive individual behavioural modes.

Thus, while science for its own sake and objectification of phenomena for its own sake stripped away the last remnants of universalized superego and rational consciousness that had been the legacy of Roman Christianity in Northern European countries, the case of Nazi Germany being the best example, national socialism as opposed to nationalism *per se* allowed for a certain amount of stabilization of collective modes within the society, if not without. The particularly precise results of science for its own sake or objectification for its own sake in Nazi Germany, leading, for at least a brief period, to a successful national socialism, was not the case when this ideology was transplanted, developed and adopted by other groups, due to the confluence of forces that differed in these other groups. In Russia, under Stalin, national socialism by no means led to a successful stabilization of internal collective modes. Vast proportions of the Soviet state in Asia possessed little or no vestiges of the 2nd efflorescence, while in the Western part of Russia much of the population was heir to stabilized forms of the 4th efflorescence, similar to societies in the tradition of Roman Christianity. Forced collectivization of peasants in Central and Asiatic Russia was meant to eradicate modes of the late stages of the 3rd efflorescence, and replace these modes with collective modes of the 4th efflorescence through revival of 2nd efflorescence behavioural patterns. In Western Russia, there was an attempt to eradicate rational, stabilized modes of the 4th efflorescence by simply eradicating those individuals in whom such modes were visibly evident. The disastrous social effects of national socialism in Russia was repeated with even more disastrous effects in China under Mao, who took Stalinism as the model by which China could be brought into collective modes of the 4th efflorescence.

The social effects in China were even more disastrous than in Russia because, except for a relatively small handful of individuals who had been educated outside of the context of Chinese society along the lines of Roman Christian society, virtually the whole population remained in the late stages of the 3rd efflorescence. By eliminating the handful of people in the Cultural Revolution who had absorbed modes of the 4th efflorescence, the desired end of national socialism in arousing collective modes of the 4th efflorescence among the population was further obfiscated. Collective will was aroused, but not the kind of collective will characteristic of modes of the 4th efflorescence, for this collective will represented that of the late stages of the 3rd efflorescence, where the authority and force of the state was both accepted and rationalized as the authority and force of the cosmic mind on the one hand, and the authority and force of cosmic libido on the other hand. In China under Mao individual consciousness manifested in constant self criticism of others reflected cosmic consciousness, while the 'practice' aspect of Maoism was an expression of libido that similarly reflected universal or cosmic libido. By attempting to eliminate the moral component of universalized superego, Maosim or national socialism in China succeeded in unleashing only aggressive libido on both the individual and the collective level, without any attenuation of these forces into rational modes.

Chaos was the result on the level of interpersonal relations, and wasteful wars on the level of international relations, while the desired ends of collective modes of the 4th efflorescence, involving rationalization of economy, society and government, were delayed for generations. In effect, national socialism in China or nationalistic or national socialistic movements in other parts of the 'underdeveloped' world more often than not represent the manifestation of hostility which civilizations who have chosen to remain in the late stages of the 3rd efflorescence feel towards those societies that have raced into the 4th efflorescence. As such, nationalist or national socialistic movements in the 'underdeveloped' countries are a manifestation of genuine cultural conflict with the 'developed' world.

Where extremist elements among the various nationalistic and national socialistic movements actually wish to engage in warfare with the civilizations of the 'developed' world, or where these movements still prevail in peripheral areas of the developed world itself, such as in the Irish Nationalist movement, techniques are borrowed from the examples of both Hitler and Stalin to create small, but extremely aggressive cadres, to engage in 'terrorist' warfare. In most parts of the 'underdeveloped' world, vestiges of the 2nd efflorescence are few, so that by necessity terrorist cadres cannot be recruited from the general population at will. Selection is made of individuals with perverse psychological problems resulting from emotionally impoverished backgrounds, or in the case of the Palestinians and other such groups where a large number of disenfranchised and often parentless individuals are available, the training of these cadres is based on principles of further personality destabilization. There is an attempt even to institutionalize such personality destabilization within the cadre. Abuse of the individuals by brutalizing parental figures over a long period of time reduces the opportunity for the individual so trained to establish an individual superego and rational consciousness. What emotional and physical gratification is allowed is identified with the brutalizing parental figure who embodies, in lieu of the absence of individual superego, the collective superego of the cadre, rationalized in turn as the superego of the national group the cadre is being trained to fight for.

The individual so trained is conditioned to obtain ego-recognition in terms of the cadre only, where all forms of naked individual libido aggression are permitted expression in the name of this kind of ego-recognition. As both consciousness, superego and ego-recognition belong entirely to the cadre as a collective, instructions to the cadre from a leader or surrogate of the original brutalizing parental figure can trigger in a seemingly automatic way almost any form of unthinkable aggressive behaviour. Not only is this the morphology of the terrorist groups who are known throughout the world today as 'liberation' movements, but also the morphology of the engineered cadres of religious cults that have been formed in the 'developed' world as a means of combating 'liberation' movements in the 'underdeveloped' world or on the periphery of the 'developed' world. In both cases where a collective superego of sorts already exists among members of the terrorist cadre, the associations of their superego are transferred to desired and often surrogate tribal and territorial associations of the cadre. Where a terrorist group is based on a religious cult, transference of an already existing superego to the surrogate religious associations of the cult are extremely effective in utilizing whatever elements of collective superego that remain extant in individual members of the cult or cadre. The phenomena of Jonestown was one such example of the latter.

The fact that deliberate creations of such barbarous evolutionary regressive forms of behaviour occurs in both the 'underdeveloped' and 'developed' world, often under the aegis of the great superpowers, is further testimony to the consequences of science for its own sake and objectification for its own sake that is part of the phenomena of modes of the 4th efflorescence. The loss of faith from both within and without in the great colonizing models of Roman Christianity is at least part of the reason for the almost ubiquitous features of 4th efflorescence societies where science for its own sake and objectification for its own sake are commonly held values. On the one hand, this problem illustrates the fragility of a stabilizing mode in the 4th efflorescence epitomized by colonizing models of Roman Christianity. On the other hand, it reveals in our time, as in the time of the ancient Greeks and Romans, the self-doubt that arises in collective modes of the 4th efflorescence when confronted with the morally superior systems in the late stages of the 3rd efflorescence. The time gap between encountering these morally superior forms of the late stages of the 3rd efflorescence and assimilating them to collective modes of the 4th efflorescence once again augurs a period of extreme peril, a period, due to modern, more sophisticated means of self-destruction available to destabilized collective modes in the 4th efflorescence in our time, threatens the very survival of humanoid development on this planet.

Even where collective modes of the 4th efflorescence are not in a state of destabilization, and where the colonizing model of Roman Christianity prevails and has in fact been refined into more modern modes, the central dilemma of the 4th efflorescence remains. In its more benign aspects, the U.S. has extended the colonizing model of Roman Christianity to a universalistic concept of capitalistic, secular democracy. Similarly, in its more benign aspects, the Soviets have extended this same model to a universalistic concept of secular democratic Communism. In the smaller and more historically stable states of Europe, this model has been blended with aspects of both the U.S. and Soviet refinements of it, in the guise of a Christian democratic socialism. Wherever any form of stabilization of this model has occurred in our time, and where both internal and external forms of libido aggression have been limited by guiding principles of universalized justice, rational consciousness, and morality, fundamental structural flaws of 4th efflorescence modes plague these societies and imperil the future of humanoid development. If collective ego-recognition and libido aggression exist as developed modes, they must have a point of expression, or an outlet.

In stabilized modes of the 4th efflorescence in our time, neither external nor internal aggression are permissible by virtue of the universalized modes of rational consciousness and superego which inhibit such aggression. Not only is institutionalized slavery no longer permitted, but also even class distinctions, as a modernized form of institutionalized slavery, are considered pejorative. Modern states of the 4th efflorescence, whether Communist, capitalist or socialist, engage in various devices to eradicate class distinctions. The redistribution of wealth and institutionalization of political, class, ethnic and gender equality are modes of eradicating class distinctions. Colonization of one country by another is similarly frowned on on the basis of the same general principles. Redistribution of wealth occurs within a given society and among various societies by international agencies who have this particular purpose and directive such as the United Nations. In smaller countries of Europe which have attempted to blend all of the current modes of the 4th efflorescence towards the ends of stability, redistribution of wealth has the added purpose of alleviating stress on individual families and thereby reducing the opportunity for inadequate parenting and personality development among the individuals of the collective. The point is that if ego-recognition and libido aggression can be mitigated on an individual level, then there will be less free flowing ego-recognition and libido aggression available to the collective ego and libido for purposes that are ultimately destructive.

Although there is no question that the more of these devices that are used, and the more thoroughly they are applied, the less will be the chances for war and aggression among and between human collectives, nonetheless the problem still remains—the collective ego and libido are extant. By virtue of their existence, a better material and social and even cultural life for the individuals of the collective is possible, but the potential for war and aggression, once there is the slightest error in this intricate system of checks and balances, is ever present and real. Furthermore, the question remains as to how collectives of the 4th efflorescence, however finely tuned and stabilized, can absorb the superior morality of civilizations in the late stages of the 3rd efflorescence without entirely losing a rational and independent collective consciousness. For the goals and values of 4th efflorescence societies are entirely at odds with the social anarchy represented by cosmic consciousness characteristic of civilizations in the late stages of the 3rd efflorescence, however such cosmic consciousness represents a superior morality.

Is it, in fact, as impossible to turn back the pages of history in abdicating collective ego-recognition and libido aggression in the 4th efflorescence as it was to turn back the pages of history in abdicating individual ego-recognition and libido aggression in the 2nd efflorescence? If this is the case, as it seems to be, then the developed world must somehow find a period of grace in which the more destructive aspects of 4th efflorescence society can be mitigated, until some new evolutionary 5th efflorescence emerges and is consolidated. Undoubtedly some way of assimilating the moral superiority of late 3rd efflorescence society will be a component of the new 5th efflorescence evolutionary development. This can be seen from the fact that since the emergence of 4th efflorescence society in ancient Greece, a stabilized form of 4th efflorescence society has always been dependent on aspects of the late 3rd efflorescence religious philosophies that could be assimilated to 4th efflorescence society without obviating its more progressive modes. The colonizing models of Roman Christianity extended in our own times in terms of French and British civilization is simply the latest result of such assimilation.

In more modern times, the attempts by all civilizations of the developed world to eradicate class differences within their own society and between the 'developed' and 'underdeveloped' worlds is the trend in this regard whose potential has not yet been exhausted. Conversely, there is the trend of reconceptualizing, in societies of the 'underdeveloped' world, their religious philosophies in terms of the collective modes of the 4th efflorescence. The abdication of aggressive wars for the purpose of institutionalized slavery, or wars inspired by what may be false fears of such, is a necessary ingredient in producing a period of grace in which mankind in his present erratic and conflicting stages of development may have the opportunity to survive and emerge into a new, and higher form of evolutionary development. Eradication of class differences among individuals of a given collective and between collectives is another method of muting individual ego and libido available for purposes of collective aggression. Above all, an attempt to utilize man's highest facility, rational consciousness, must be made, so that the frustrations, fears and anxieties which members of different collectives have towards one another will be alleviated, thereby again denying free ego and libido forces to the purposes of collective aggression. Through rational consciousness, mankind can be made to understand the nature of his evolutionary genesis, and to see himself in the kind of perspective that will bind his ego and libido, both individually and collectively, to evolutionarily constructive rather than destructive purposes.

Chapter VIII

THE EGYPTIAN EMPIRE ESTABLISHES THE BASES OF ALL SUBSEQUENT HUMAN CIVILIZATIONS

Since science is an early rather than a late form of humanoid evolution, mankind has possessed the scientific wherewithal to understand himself from an evolutionary perspective since the 2nd efflorescence. During the late and middle phases of the 3rd efflorescence, this scientific capacity was used to a very full extent in explaining man's evolutionary genesis, and the debate which centered about his own evolutionary identity was the focal point at which cultures diverged, some deciding to remain in the late stages of the 3rd efflorescence, others racing ahead into the 4th. Some cultures, such as those of the Far East, made firm decisions about remaining in the late stages of the 3rd efflorescence and consolidating the gains made in this evolutionary phase for the purposes of collective survival, while others such as those in the Near East hovered indecisively between the late stages of the 3rd efflorescence and the early stages of the 4th.

In Mediterranean civilization, epitomized by the early Greeks, the decision was made to come full fledged into the 4th efflorescence, while assimilating and absorbing certain of the morally superior attributes of the late stages of the 3rd efflorescence they were constantly exposed to in neighboring Near Eastern civilizations. In Nordic civilization, conditioned by extreme climatic and ecological circumstances, and with little or no contact with civilizations of the late stages of the 3rd efflorescence until the Roman period, collectives leaped from 2nd efflorescence society directly into the 4th. In other parts of the world, and under more favourable climatic and ecological circumstances, such as Africa, the Americas, Polynesia and Australia, where humanoid collectives were for the most part largely untouched by contact with civilizations in either the late stages of the 3rd efflorescence or the 4th, humanoid collectives remained largely in the early or middle phases of 3rd efflorescence evolutionary development, with variable ill-defined modes of the late stages of the 3rd and the 4th where occasional contact with these kinds of societies were influential.

The earliest societies to have peaked the 3rd efflorescence were those in the Near East and this was due to a number of factors. Favourable climatic and ecological conditions in the Nile valley allowed the wide-spread and extremely successful profusion of agricultural techniques learned in the middle stages of the 3rd efflorescence. In addition, the seemingly unlimited pool of slave labor available from Africa permitted institutionalized slavery on a scale hardly precedented in history, either before or after the emergence of the Egyptian Empire 20,000 years ago. This pool of African labor was drawn from generally docile tribes entrenched in the passive phases of early 3rd efflorescence development. While these African collectives had just recently entered 3rd efflorescence society, climatic and ecological conditions in Africa were such that it is likely that emergence from 1st to 3rd efflorescence was rapid. Environmental conditions in Africa had not favoured entrenchment of 2nd efflorescence society, for as soon as one group emerged into 2nd efflorescence it always found almost immediately available groups of the 1st efflorescence who could be employed as slaves, such slavery hastening the ascent of the 1st group into 3rd efflorescence. Once the African group was thus entrenched in 3rd efflorescence, the group from the 1st efflorescence, by virtue of the moral imperatives of institutionalized superego, would be quickly brought up to the same level as the enslaving group. In Africa itself, however, ecological conditions were not favourable enough to permit the widespread promulgation of agricultural methods learned in the 3rd efflorescence such as occurred in the Nile valley.

When 3rd efflorescence society peaked in the Nile valley sometime between 20-50,000 years ago, the collectives of that region found a willing and docile source of slave labor for extension of their agricultural systems that veered between 1st and 3rd efflorescence development. The building of cities began, a ruling group was formed, and all the accoutrements of civilization developed in 3rd efflorescence society such as architecture, cooking, pharmacy, medicine, engineering and religion were extended over larger territories and attained the status of institutionalized modes of social organization. Due to the open geographical terrain of the Nile valley region, and the enormous wealth and surplus of agricultural production that developed there, this society, being still in itself of a rather docile nature, was succulent prey for the aggressive and free-roaming and probably free-sailing hordes of Nordic civilization who eventually conquered it. Thus, at its base early Egyptian civilization was produced by collectives who had peaked into the 3rd efflorescence and who held enslaved collectives who veered between 1st and early 3rd efflorescence from Africa. The superstructure was composed of stray collectives of Nordic civilization who had fused 2nd and 4th efflorescence modes without ever having become entrenched in 3rd efflorescence development. This melange produced the extraordinarily rigid class and slave society of ancient civilization whose legacy would ultimately be the societies of the late stages of the 3rd efflorescence and the 4th that continue to exist into the modern world.

As the Egyptian empire beginning about 20,000 years ago produced prodigious scientific marvels which even today we little comprehend, the basic difference between the kind of scientifically oriented society that existed then and the kind of scientifically oriented society that exists now, particularly in the 'developed' world, is that the ancient Egyptians still had no written, and probably no spoken language, that provided for the objectification of phenomena. The remnants of written Egyptian script that are still extant today and which are from 12-20,000 years old reveal a written and probably spoken mode of thought and communication that is almost entirely symbolic. In this subjectification of phenomena that Egyptian hieroglyphics reveal, there is little differentiation between thought that is mathematical in intent, and thought that is social, psychological, political and religious in intent. Egyptian hieroglyphics far more resemble modern computer codes or computer machine language than they resemble the memorizable string of particularized symbols common to languages of both the late stages of the 3rd efflorescence or those of the 4th. There are no phonetic components of Egyptian hieroglyphics as in Indo-European languages, nor are there any components that are identifiable by visual association rather than visual representation as in modern Chinese which emerged 5-10,000 years ago.

Modern Semitic languages are composed of a combination of the latter two, containing some components that are phonetic and some that operate as a result of visual association. The components of the hieroglyphics cannot be broken down and their meaning changed by recombination of symbols as is true in both modern Indo-European, Chinese and Semitic languages. The meaning of the symbols of the Egyptian hieroglyphics, containing thus no grammar of any recognizable form, cannot be induced from the context of a string of such symbols. Thus Egyptian hieroglyphics, despite the most intense efforts in decoding, remained entirely untranslatable to modern man until the discovery of the Rosetta stone early in this century, where a text of ancient Greek existed as an exact translation of the hieroglyphic text on the stone.

It is above all this lack of intra-contextual significance which separates Egyptian hieroglyphics from all modern language forms. It would be necessary to first know the subject to which the hieroglyphics refer, step by step, thought by thought, before they could become intelligible. If many tens of millenia from now some new race were to discover modern computer machine codes of modern 'developed' society, they would have the same problem deciphering its meaning as we have of deciphering ancient Egyptian hieroglyphics. The amount of information which each computer symbol designates cannot be inferred from the computer symbol itself or even from the context of a number of computer symbols. This information has been learned prior to the setting down of the code and has been learned in a language and a mode of thought that bears little relation to that code and its symbols. Thus, as in modern computer codes, the symbols of Egyptian hieroglyphics have referral or reference functions, rather than explicative or cognitive functions *per se*. In addition, the body of information to which these symbols refer were learned, understood, and ultimately collated in another manner, probably of a mathematical nature.

The extraordinary engineering and pharmacological feats of the Egyptians which, even by modern standards to all intents and purposes remain inexplicable, strongly suggests that this would have been the case. Even assuming the massive input of slave labor over a number of generations, modern engineers can still not explain the scientific and engineering aspects of pyramid construction. The ability to preserve human flesh in more or less extant form for 20,000 years is far beyond the ability of modern medical or pharmacological science, yet in both of these cases it was science and not magic, which was responsible for these feats, the existing evidence of which is incontrovertible.

As in any modern master computer code, where more than one body of information may be represented by a common number of symbols with key reference symbols or addresses designating which body of information is the particular subject of a particular message, so in Egyptian hieroglyphics the same methodology seems to have been applied. This is one explanation for the relatively limited number of symbols in Egyptian hieroglyphics, where in any given message key symbols or addresses will differ where the majority of the other symbols will remain the same. Given the fact that in this period of humanoid evolution, information and communication was expressed by subjectification rather than objectification of phenomena, all fields of cultural interest and investigation were blended in a continuum, so that there was no clear, no objective dividing line between science, art, religion, magic, social organization, politics, medicine, pharmacy, philosophy, etc. The master computer-like code which Egyptian hieroglyphics represents therefore is an expression of such a continuum, where specific symbols in a given message designate the particular field or fields of information that are addressed.

By the time the Greek translation of the Egyptian text on the Rosetta stone was written, about 2500 years ago, 15,000 years or more had passed since the consolidation of the Egyptian Empire. By this time, and in contact at a very late and decadent stage of the Egyptian Empire, the computer-like code of Egyptian hieroglyhpics had been adjusted to refer to the meaning components of a modern language such as Greek. The fact that there exists this single example of such adjustment is testimony to the difficulty and inconvenience that such adjustment implied, and in fact as far as can be seen, Egyptian hieroglyphics as a written mode disappeared at the same time as the Egyptian Empire entered its final phase about 3000 or 4000 years ago. Given the rigid class and slave structure of the Egyptian Empire, it is clear that mastery and understanding of Egyptian hieroglyphics were the province of an elite class and were never used or properly comprehended by the masses of the Empire. When the Egyptian Empire dissolved about 3-4000 years ago, the class and slave structure was also in a state of dissolution, and in an attempt to find a written mode to express their thoughts, Egyptians then adopted the Semitic type languages from neighboring civilizations that were in fact within the modern gender of language.

The religious, social and political system of the ancient Egyptian Empire was such that the entire energies of the vast population under Egyptian control were employed in the service and in the worship of the Pharaoh of Egypt and his immediate family. This service and worship included not only contributing economically, socially, and politically to the welfare of the Pharaoh and his immediate family, but also contributing to those spectacular engineering and pharmacological feats that were meant to bring about the reincarnation of the Pharaoh and some members of his immediate family. Never in the history of humanoid development, neither before or since the ascendancy of the Egyptian Empire, has there existed a situation where so many individuals were so thoroughly oppressed in the interests of so few. Even the most tyrannical regimes known in modern times pale by comparison and seem positively egalitarian compared to the tyrannical rule of the Egyptian Empire. The Pharaoh was conceived of as a living God, and as such all the energies of the populace were mobilized to serve him, not only in his lifetime but also after his death. The prodigious and spectacular scientific achievements of the Egyptian empire were not utilized for the collective in any way, but only the living conditions of the Pharaoh, and even more so, to improve the conditions in death for the Pharaoh. The scientific marvels of the Egyptian tombs indicate that life after death for the Pharaoh was in fact far more important than the comforts the Pharaoh enjoyed in life, since there are no palaces extant for Pharaohs during their lifetimes which even begin to match the splendour of the domiciles planned for them in death.

In the Egyptian Empire, collective ego-recognition and libido aggression reached an extreme unknown in societies since, in societies that exist or existed in the 4th efflorescence proper. Even societies in the 4th efflorescence in modern times which contained powerful vestiges of the 2nd efflorescence, such as Nazi Germany or Stalin's Russia, did not begin to approach this extreme, ameliorated as they were by historical experience since the time of the Egyptian

Empire and which experience is in no small measure part of the legacy of the Egyptian Empire, its vicissitudes and its aftermaths. For the Egyptian Empire was the dawning of collective modes of the 4th efflorescence in the context of unbridled forms of the 2nd, and with all the skills and capacities of the 3rd in its peak at its disposal. Institutionalized superego at its peak, having produced all of the accoutrements of civilization, attained a consciousness of collective self that was essentially moral. Institutionalized superego and collective consciousness was free-ranging in this peak period, and embodied itself in all aspects of nature and man, which allowed for development of the accoutrements of civilization in which this moral sense was infused in a continuous way.

The Nordic invaders and conquerors of the Nile valley however had not attained this moral sensibility. They were in a phase that fully expressed the modes of the 2nd efflorescence with vague, ill-formed institutionalized superego that was limited to the small collectives in which they roamed together, pillaging and conquering others as an expression of their own individual ego-recognition and libido aggression. The success of cumulative ego-recognition and libido aggression in a collective banditry led to an embryonic sense of 4th efflorescence modes of collective ego-recognition and libido aggression. Once they conquered the people of the Nile Valley, they found they had to roam no further, for pillaging was institutionalized as a mode of class and slave-state exploitation. The conquered people, however, having peaked into 3rd efflorescence society immediately identified their aggressive and self-confident conquerors as the moral force their own well-institutionalized superego had sought. They consolidated the range of moral and supernatural sensibility in the willing persons of their conquerors. This symbiotic relationship quickly led the Nordic conquerors of the Nile valley to reinforce the fusion of 2nd and 4th efflorescence modes that they had brought with them as part of their own ecological and climatic conditioning under unfavourable environmental circumstances. The concept of themselves as moral and supernatural beings which the people they conquered forced upon them bolstered their sense of both individual and collective ego-recognition and libido aggression to unimaginable proportions. As beings who thought of themselves as Gods, and who yet had no real notion of the moral or supernatural forces of the universe, they became unwilling to accept the idea of their own death.

The ultimate termination of ego-recognition and libido aggression in death was the dilemma that 2nd efflorescence development had brought about in humanoid evolution and which endangered the very survival of humanoid evolution in the first instance. In the general trend of humanoid evolution, it was the progression from 2nd to 3rd efflorescence which ameliorated this dilemma and ensured, at least until the development of 4th efflorescence modes, the survival of humanoid collectives. The Nordic conquerors of the Nile valley had had at that time little first hand experience with 3rd efflorescence modes, so that the adoration shown to them by the people they conquered, on the basis of their own institutionalized superego, only enhanced their own ego-recognition and libido expression, making the idea of death even more painful and unbearable than it had previously been.

In the general trend of evolutionary development, enslavement of others by humanoids in the 2nd efflorescene normally meant that those who were enslaved were in an earlier period of efflorescence than the enslavers. Under these conditions, institutionalized slavery meant that the enslaving group had an opportunity to unleash naked ego-recognition and libido aggression on an unwilling and terrified slave collective, allowing free expression of brutality and hostility towards the enslaved group that might otherwise have been unleashed upon the relatively helpless females and offspring of the enslaving group. This provided a measure of safety and security for females and offspring which led eventually to the full development of the 'female type' and institutionalized superego in a 3rd efflorescence phase of evolutionary development. However, in the case of the invaders of the Nile valley, the enslaved group was in a later period of efflorescence than the enslavers. When brutality and hostility were unleashed against this enslaved group, they were not terrified, for they willingly accepted such brutalization as the will of the supreme moral force of the universe which they recognized in the persons of their conquerors. Such willing acceptance of brutalization had the obverse psychological effect from unwilling and terrified resistance to brutalization.

The willing acceptance of brutalization on the part of the enslaved population provided no real emotional outlet for aggression and hostility on the part of the enslavers, for the willingness with which brutality was accepted and even desired allowed for no nemesis or self-recognition on the part of the brutalizing enslavers that would have assisted them with superego development. Instead it allowed the enslavers to internalize their brutal modes of behaviour as an aspect of ego-recognition and thereby enhance it. Libido-aggression became a substitute or surrogate for superego, as it was given a rational, almost moral kind of value in terms of ego-recognition. The conquerors of the Nile valley learned to place a rational and moral value on their own libido aggression, so that both rational consciousness and moral sensibility developed in a pathological way.

The concept of a universal moral force was internalized in their own egos. This kind of pathological personality development resulted to some extent in benevolent forms of rule on the part of the conquerors; since it was necessary to maintain the enslaved group to a large extent intact, the enslaved populace had become internalized in the psyche of the rulers as a means of validating their own ego-recognition and libido aggression. In addition, the spectacular wealth and the diverse accoutrements of civilization possessed by the enslaved group was of direct benefit to the conquerors, both in terms of the luxurious way of life it permitted them to enjoy and also in terms of enhancing their own expression of ego-recognition and libido aggression. This was all the more the case since the means by which such accoutrements of civilization had been achieved by the enslaved group remained inaccessible to the enslavers, who themselves had not passed through the peak of 3rd efflorescence development that would have made the means of acquiring such accoutrements a natural mode of their behaviour and understanding.

Before the conquest of the people of the Nile valley by Nordic hordes, cultural modes of the peak of the 3rd efflorescence were diffused among the general populace with particular refinements of this society residing in certain specialized roles which more or less embodied or symbolized certain features of 3rd efflorescence culture, such as in priests, doctors, architects, artists, agronomists, engineers, etc. On the whole, however, knowledge of all parts of the cultural achievements of middle 3rd efflorescence society was diffused, so that the transmission of particular aspects of the culture was not determined by rules of social organization, such as caste, class or guild. Like the culture itself, conceptualized and understood in terms of subjectification of phenomena, social organization was diffuse and free-ranging, so that the son of a farmer in one generation might be a doctor in the next, as the spirit moved the individual and as society recognized such synchronistic and eclectic movement of talent and role as an aspect of the moral and supernatural force on which the foundations of their entire culture and society were built. However, with the advent of their conquest by Nordic hordes, this situation changed. The new rulers of these people of the Nile valley were seen to embody this moral and supernatural force which was in fact the product of their own institutionalized superego and which was the foundation of their social and cultural modes. Their collective sense became focused. This focus, indeed, was all the more pointed by the particular psychological makeup of the ruling conquerors who themselves reinforced this collective sense of the conquered peoples in terms of their own individual and collective ego-recognition and libido aggression and reimposed this conception upon the docile people whom they had conquered.

The result was that the cultural and scientific achievements of the people of the Nile valley for the first time became both directed and purposeful, organized towards the specific ends of enhancing the ego-recognition and libido aggression of the rulers in whom was embodied the collective sensibilities of both conquerors and conquered, albeit with somewhat markedly different emotional affects in these two distinct groups. Roles that had previously been diffuse were structured according to a melange of systems that included class, caste and guild. This was necessitated by the fact that the rulers demanded a means by which they could have access to the cultural and scientific wealth of the conquered peoples, a means which of necessity obviated their having any direct access to the cultural modes that made their wealth possible.

Thus, a form of political control over this wealth was institutionalized, which necessitated that the social organization of the conquered peoples became rigidly organized and structured. Where the wealth of cultural and scientific knowledge existed, hierarchical patterns were imposed, so that in any given field of cultural or scientific knowledge, resources and talents were organized and pooled with a leader of the individuals possessing a particular cultural or scientific skill answerable to the conquering elite. The conquering elite did not want to know the intricate logic behind the accomplishments in any given cultural or scientific field; they merely wanted to use these accomplishments for their own ends of ego-recognition and libido aggression, collective and individual. The politically designated leader from any particular field of cultural or scientific knowledge was of interest to the ruling elite only insofar as he could make the knowledge involved available to the rulers for their own ends of ego-recognition and libido aggression. If he did not satisfy the ruling elite in this regard, he was dispensable, for from the point of view of the ruling elite all of the conquered people were seen as slaves to be dealt with at the whim of their most brutal and hostile libido aggressive expression. Given the peril in which politically designated leaders from any particular field of cultural or scientific accomplishment lived and died, it was necessary to have an ordered system of succession within any given field, so that with the execution or incapacitation of any leader who displeased the ruling elite, another was ready to take his place.

These kinds of circumstances led to an organization of resources, role and talent in any given field of knowledge on the basis of class, caste and guild, an organization that was seen from the point of view of the conquered people as an aspect of the moral force of the universe as embodied in the ruling conquerors. From the point of view of the conquerors, it was seen as politically necessary evolution that allowed them to maintain the enhanced sense of ego-recognition and libido aggression they had learned to enjoy as they became accustomed to integrating moral imperatives into their own sense of ego-recognition and libido aggression, individual and collective.

Evolution of social and political modes among the conquered peoples, coupled with the necessity of organizing the cultural and scientific wealth of their society into directed and purposeful ends, no doubt contributed to a further evolution in the body of scientific and cultural accomplishments which they already possessed. The nemesis or self-recognition of their brutalization by the ruling elite, although continuing to be seen in the light of a moral and supernatural force, allowed them to enhance their sense of cosmic morality to include a sense of cosmic immorality or amorality. They began to develop a sense of cosmic libido and cosmic consciousness, which in time and under favourable circumstances they were able to separate from cosmic superego or moral force. In the earliest Egyptian myths, Ra, the sun God, has a total moral force whose machinations for good or for evil, were undistinguished. In later Egyptian myths the Ra God-figure exists concurrently with the Osiris God-figure who is in fact often seen as another form of Ra or indistinguishable from him. Osiris-Ra is a God who is not only powerful, but also good, a manifestation of cosmic morality as distinguished from cosmic immorality. Osiris is slain by his evil brother-in-law, cousin, Seth, who represents universal libido and is characterized as King of the underworld or hell, while Osiris is saved from death by his wife, the Goddess Isis, who with her extraordinary magical, scientific and religious powers is able to reassemble the parts of his body that were dismembered by Seth. She is however unable to restore his masculine organs, and the two of them reign in the upper world of heaven and earth with their infant son Horace, conceived and born before Osiris's dismemberment.

By the time of the Osiris myths in the early-middle period of the Egyptian Empire some 15,000-20,000 years ago, collective consciousness among all segments of Egyptian society had been extended to include a sense of cosmic libido and cosmic consciousness clearly distinguishable from cosmic super-ego. Isis represents the deification of the 'female type' whose entrenchment in Egyptian society made possible this distinction on the moral plane between good and evil in both social and cosmic forces. The loss of Osiris's masculinity makes him representative of cosmic consciousness, as opposed to the cosmic libido represented by Seth. The loss of his masculinity deprives Osiris not only of libido, but also of ego, on an individual and collective level, so that collective consciousness no longer comprises collective ego-recognition and libido aggression as its major components. The infant Horace who by virtue of his age is also devoid of ego-recognition and libido aggression represents cosmic morality or superego.

In another myth of this period, Isis poisons the God Ra and through the mediation of a snake at her command threatens to let him die if he does not reveal to her his real name or identity. When he gives in and does so, she saves him by virtue of her legendary religious and scientific powers. Here again we have an even clearer indication that the cosmic power of the universe, as embodied in the God-figure, Ra, relinquishes his ego-recognition to the 'female type' represented by Isis in order to maintain his God-like status. In all of these myths, the female Goddess Isis, personification of the 'female type,' represents the institutionalized superego of the middle 3rd efflorescence society, as she is identified so closely with the religious, scientific, medical, linguistic and pharmacological skills that were the cultural achievements of this phase of humanoid evolution. Her acquisition, at least in part, of the ego-recognition which she steals from the Gods, embodying cosmic forces of a free and rather indistinguishable nature, reflects the evolution of consciousness in Egyptian society under the influence of conflicting but always interacting modes of evolutionary development.

That this evolution of consciousness in Egyptian society as reflected in these myths had little opportunity to become entrenched in more liberalized social and political modes is a matter of historical record. So extreme were the differences in collective modes between the rulers and the ruled in Egyptian society that amelioration of the oppressiveness of the regime proved impossible. It was only in the much later diffusion of this evolved form of consciousness to neighboring mid-eastern collectives, or in the rebellion and exodus from the Empire itself of collectives, such as the Hebrews, that this evolved form of consciousness was able to produce a less oppressive form of political and social structure. In Egyptian society itself, while where for limited periods this evolved form of consciousness might have prevailed upon the tyrannical and oppressive regime that existed there, on the whole the myths which arose from this new evolved form of consciousness were fused with older myths and forms of consciousness that entrenched and in fact helped to expand and extend the existing social system. While the ameliorating social and political ramifications of this newly evolved consciousness were subsumed under the older melange of collective modes, in fact they reflected an advance in scientific and cultural achievements which were separated from the social implications of these achievements, and utilized to further the aims of collective and individual ego-recognition and libido aggression of the ruling elite.

Since all knowledge in this period was conceptualized in terms of subjectification of phenomena, an advance in one field of knowledge necessarily meant an advance in all fields, as phenomena, both external and internal, was seen in a continuum where no clear line was drawn between knowledge of, for example, medicine, or religion, or social organization, or science. The evolved ability to make distinctions between good and evil, or superego and libido, or ego and consciousness, heralded similar spectacular advances in science, engineering, medicine, and pharmacology. The religious ramifications of such evolved consciousness, rather than being extended to the social and political sphere, were utilized solely in the scientific and technical spheres which served the particular ends of ego-recognition and libido aggression on the part of the Pharaohs and their immediate families and ruling cliques. The end towards which all this new knowledge was directed and organized remained the same as when technical and scientific knowledge was more limited: towards preventing the termination of ego-recognition and libido aggression for the Pharaohs and their immediate families and favorites. With the advances in scientific and technical knowledge, the ruling elite became more determined than ever to not only have life after death, but also to be reborn again, in the same shape and with the same ego identity that they had possessed in this life and with which they were so loathe to part.

Chapter IX

DISPARITY IN SOCIAL AND TECHNOLOGICAL ADVANCES IN THE EGYPTIAN EMPIRE

From the point of view of 4th efflorescence society in modern times, it seems incredible that the entire resources and energies of a population, over a period of nearly 20,000 years, could have been directed and organized towards the bizarre ends of ensuring the reincarnation of its leaders. Yet, the historical evidence is incontrovertible. The duration of this social and religious excercise of 20,000 years is 5 times the time period which we designate as the 'ancient' world. In this short period since the final collapse of Roman civilization, all the modern nations of the 'developed' world were formed, none of them having been in existence before 2000 years ago. Even the great civilizations of the 'underdeveloped' world which remain in the late stages of the 3rd efflorescence had their origins in proto-civilizations that are not older than 5-10,000 years old, such as the Babylonian, Sumerian, Persian, Indian, Hebrew and Chinese civilizations, and in every case the genesis of civilizations that today remain in the late stages of the 3rd efflorescence originally occurred as some form or other of a reaction to Egyptian civilization and its empire. Whether or not reincarnation is a scientific possibility remains an irrelevant question from the point of view of the best scientific and philosophical thinking of 4th efflorescence society, although in most of central, south and eastern Asia, as in parts of North Africa, there is more than a lingering faith that reincarnation may indeed occur. Nowhere, however, in the 'underdeveloped' world, where ego-negating and libido abdicating philosophies prevail, is reincarnation seen in the same light as it was under the ancient Egyptians. For nowhere today does belief in reincarnation include the belief that the reincarnated individual will be reborn again in exactly the same shape and form and with the same ego-identity in another lifetime as he had in a previous lifetime.

From a modern scientific point of view, it is possible, if not probable, that the permutations and combinations of atomic, molecular, and biochemical structure which determine the certain shape and form of a given individual would occur more than once. Even in the unlikely circumstances of such an occurrence from a scientific point of view, there is no way to even conceptualize the possibility of an individual with identical chemical and biochemical structures in two points of time having an identical ego identity. In Asia, where belief in reincarnation persists, this problem is bypassed by the belief that reincarnation has neither necessary biochemical similarities nor ego-identity replication, but does have the attributes of common individual consciousness, this individual consciousness being inextricably linked with the cosmic consciousness that guides the workings of the entire universe.

The theory is that when individual consciousness is properly attuned to cosmic consciousness through methods that enhance the individual's ability to subjectify all phenomena, he may be able to recall 'himself' or individuals similar to 'himself' in previous lifetimes or even in concurrent space-time coordinates. Such recall, from a modern point of view, is an aspect of the conclusions implied by the work of Carl Jung, that given the similar circumstances of evolutionary development among humanoid collectives, thought itself evolves concurrently with biological or bio-psychological evolution. As every individual passes through the phases from conception to adulthood, the history of the race is repeated in his individual development, and given an adequate amount of information about human history and more than adequate amount of imagination, there is no reason why any individual could not imagine what he would have been like, or would have thought or done, had he lived in another period of time.

Of course, the problem is really not that simple, as either any psychiatrist or Orientalist will testify. In an era when objectification of phenomena is almost universally the norm, at least in the 'developed' world, the ability to subjectify phenomena may be the mark of great talent, or an example of tapping powerful spiritual resources, or on the other hand may mark the transition from mere neurosis to full-blown insanity. More often than not, at least in the developed world, the latter is the case, and mental hospitals are full of people who believe they are living the lives of beings long dead, or who have a multitude of ego-identities, all of which seem to them to be equally real. These pathologies are invariably associated with inadequate personality formation in early childhood; such pathology is created by the fact that the individual is fixated in an evolutionary developmental stage that is contrary to the norm of the society in which he actually lives. When, by virtue of poor personality development, the individual is fixated in a mode prior to the 4th efflorescence, he necessarily has proclivities for subjectification of phenomena.

The conflict which results from these proclivities coupled with the objectification of phenomena modes that are imposed on him in later developmental stages by the society in which he lives may give rise to such fixations, fixations that are the objectification of ego identities in a prior lifetime or ego-identities existing simultaneously in a concurrent lifetime. The methods that Freud developed to unwind these convoluted pathologies involved the guided self-examination of the individual's own semantics, the semantics in which he had learned to conceptualize himself in a mixture of subjectification and objectification of phenomena. Where Freud's lengthy curative methods of psychoanalysis have failed, the modern methods of combining drugs or shock-treatment with psychotherapy are more successful. The effect of the biochemical compounds and shock-treatments administered to the mentally ill have the effect of dampening cognitive processes, or the neural biochemistry which produces them, so that convoluted objectification of phenomena in which subjectification of phenomena is rationalized to the self is muted to the point where examination of the individual's verbal semantical conceptualization of himself can be successfully dealt with in psychotherapy.

Dampening of the cognitive processes in mentally ill individuals is the method by which the pathological 'life adjustment' or self-rationalizations of the individual may be obviated to the point where the causes rather than the symptoms of disease may be treated. The causes are invariably rooted in poor psychological development patterns that occurred in early childhood, and where the imbalances in developing ego, superego and libido resulted in modes of subjectification of phenomena. Rational consciousness was thus unable to develop in a proper way, so that objectification of thought, which is an evolutionary product of rational consciousness, developed in pathological ways. On the other hand, in the 'underdeveloped' world where ego-negating and libido abdicating religious philosophies prevail, mental illness as such is almost unknown. The most direct cause for the lack of mental illness of this kind is the socially and culturally enforced mitigation of objectification of phenomena. Thus, the convoluted and complicated rationalizations and 'life adjustments' in modes of behaviour that mark the mentally ill person in 'developed' societies have little or no chance of occurring.

Where the individual from the late stages of the 3rd efflorescence believes himself to be a reincarnation of individuals who lived in previous times or believes himself to have a number of ego-identities in concurrent time-space coordinates, this is conceptualized by him in terms of a coherent religious philosophy around which his personality is organized and which presents no contradictions in social intercourse. Rational consciousness, as opposed to individual consciousness, in such societies is discouraged anyhow, so that the belief in reincarnation by the individual is not judged, and thereby contradicted, by rational standards. The sense of reincarnation is associated with the super-rational cosmic consciousness which the individual is encouraged to acquire, and by standards of cosmic consciousness, a sense of reincarnation is entirely logical. Attuned to cosmic consciousness, the individual can flow with the universe and be at all times and all places, even at the same time and in the same place.

If we were to evaluate the psychological makeup of individuals in the Egyptian Empire of 10-20,000 years ago, we would have to conclude that their sense of reincarnation placed them somewhere between the two extremes of personality types that exist in the modern world, and believe in the reality of reincarnation. They veered between the prototypes of the severely mentally ill individual of the 'developed' world who believes in reincarnation and the faithfully religious and entirely normal individual in the 'underdeveloped' world who believes in reincarnation. For the mass of the Egyptian people who remained in a state of institutionalized and total slavery, the psychological makeup of the individual who believed in reincarnation would have more closely resembled present day believers in reincarnation in the 'underdeveloped' world and who remain psychologically normal, stable and adjusted. For those members of the Egyptian masses who became a political, social and technical elite among their own people, and for the ruling Pharaohs themselves, the psychological makeup of the individuals who believed in reincarnation would have resembled pathological individuals in the 'developed' societies of today who believe in reincarnation.

The Nordic conquerors who formed the Pharaonic elite of Egyptian society had had little experience with the middle stages of 3rd efflorescence society and had, at best, ill-formed institutionalized superego. Subjectification of phenomena on any level that would have enabled them to integrate all the dynamics of individual personality development was extremely unlikely. Confronted with a society that enabled them to enhance their own ego-recognition and libido aggression through the wealth of scientific and cultural achievements that were the product of peak middle 3rd efflorescence evolutionary development, they would, of necessity, have been inclined to conceptualize this enhancement by means of objectification of phenomena, however crude and primitive this objectification might seem by modern standards where such objectification has evolved from the development of rational consciousness. Similarly, the politically designated elite of the Egyptian masses, having been forced to organize the scientific and cultural achievements of their society to a directed and purposeful end, would likewise have been motivated to conceptualize in terms of objectification of phenomena, since in fact there would have been no other way to organize this wealth of information to a directed and purposeful end, as well as no way to communicate the effect of this information to their masters in a way that they could understand and be satisfied with. For the political elite of the Egyptian masses, development of a way of conceptualizing information in terms of objectification of phenomena, again, however crude and primitive by modern standards, was a matter of sheer survival.

What precisely were the kinds of technical, scientific, engineering and medical accomplishments and accoutrements of the ancient Egyptian Empire is again a subject which must be tackled from the point of view of the social and political circumstances under which this empire existed and continued to survive for so long. Without overstating the facts, the historical evidence is such that we must assume that the Egyptian empire possessed scientific and technical achievements and accoutrements at least equal and in many cases superior to those developed by modern 4th efflorescence society and accessible today to all peoples of the world. That the actual mechanical engines and chemical laboratories which must have existed in ancient Egypt necessary to produce the construction of sheer walled pyramids and other marvels of engineering, and to preserve human flesh almost intact for 20,000 years, are no longer extant is no indication that they did not at one time exist. The knowledge of electricity, for example, is known to many societies who have not ever passed the middle phase of the 3rd efflorescence, and in our own time knowledge of electricity only became disseminated because of the social uses it possessed. In liberalized societies of the 4th efflorescence in modern times all technical and scientific advances have been regarded as the province of all individuals in the collective. This was most certainly not the case in the ancient Egyptian empire, where scientific and technological advances were utilized for the sole purpose of the Pharaohs, not even so much for ameliorating their life on this earth, but for increasing their possibility of being reincarnated after death, for in fact obviating the loathsome and painful termination of ego-identity and libido aggression.

In each generation of the Egyptian empire, all scientific and technological advances and skills were utilized for the bizarre and impossible purpose of eliminating death for a single individual, the Pharaoh, and for his Queen, and perhaps one or two other favorites whom he obsessively wished to spare from the same unthinkable end, as his own God-like existence was destined to meet. Thus as electricity must have been known to the individuals of the peak middle 3rd efflorescence society of ancient Egypt, utilization of electricity only occurred under circumstances where scientific and technical knowledge was organized for the limited purposes of reincarnation. Utilization of electricity in mechanical and electromechanical devices under the circumstances of the Egyptian empire would have been under the purview of a very few, certain members of the political elite of the Egyptian masses and perhaps the Pharaonic clique.

Undeniably, electricity, and other means of reducing human labor input in the construction of pyramids and other structures central to the purpose of the empire, would never have been used in order to ameliorate the working conditions of the enslaved masses who were employed in building these structures, but only in accomplishing engineering feats which could not have otherwise been accomplished by human labor, however difficult, alone. In addition, in the Pharaonic collective where paranoia would have been a 'normal' aspect of individual and collective behavioural modes, focued as these modes were on the ever-increasing enhancement of ego-recognition and libido aggression, any technical inventions that might have ameliorated the living conditions of the masses were certainly kept secret from them, and no doubt great pains were taken to ensure the inaccessibility of such devices to the Egyptian masses. Under such circumstances where the number of such devices were of necessity extremely limited and where great pains were taken to ensure information about them was not disseminated among the people, it is not surprising that after a period of 10,000 years, which augured both the peak period of the Egyptian empire and the moment of its decline, that such devices have not been discovered intact among the Egyptian ruins and burial artifacts which are the only remaining record of this civilization.

From indirect evidence of the ancient Greek period about 2500 years ago we know that mathematics, geometry, and trigonometry, put into its present form of objectification of phenomena, was derived from the Egyptian empire, where the knowledge of these subjects straddled the forms of subjectification and objectification of phenomena. Geometry and trigonometry, which have played such a vital role in the development of science and technology in modern 4th efflorescence society, still remain useful only insofar as they are graphically and visually representable, illustrating the vital importance of subjectification of phenomena in any sphere of science or mathematics, irrespective of the phase of humanoid efflorescence in which it appears. The single most important mathematical formula, in terms of its significance for the development of mathematics and science in modern 4th efflorescence society, is the theorem of Pythagoras, which algebraically models the structure of a right-angle triangle. Pythagoras' mathematical work, according to ancient Greek history, arose out of a religious cult with which he was associated, and which had its roots in the ancient religion of the Egyptian empire. Similarly the binary counting systems of the ancient Babylonians which appeared about 5-7,000 years ago, and which is similar to the binary counting system of the ancient Chinese as found in the Chinese *I-Ching* at about the same time at least and probably earlier, also appears to have been derived from contact with the mathematics and technologies of the ancient Egyptian empire. Of course, it goes without saying that construction of the pyramids and other such engineering marvels that are found among the ruins of ancient Egyptian civilization could never have been constructed without extraordinary and complicated mathematical systems.

The abacus, which is a mechanical counting device or calculator that utilizes both the binary and decimal system still used today in the modern Near East and Far East, also would appear to have its origins in the ancient Egyptian empire. That the Egyptian empire utilized electro-mechanical devices for calculation or possessed the equivalent of modern day computers, is also probable considering the feats they were able to accomplish in the realms of engineering and medical chemistry. Egyptian hieroglyphics resemble nothing so much as modern day machine-language codes for computers, and the long messages left on any given royal Egyptian ruin may have been important, not merely as a religious memorial, but also at the same time as the machine-language code signifying or marking the particular computer programme utilized in the construction of that particular monument.

If all of this may seem far-fetched to individuals in modern 4th efflorescence society, it is because they have been conditioned to believe, by virtue of their somewhat limited collective experience and the collective ego associated with this experience, that technical advances such as electrical, electro-mechanical devices and computers are the result of a spirit of experimentation that is peculiar only to them, never having existed prior to their own emergence on the historical stage. Nonetheless, it is a matter of historical record that the achievements in medical science, and pharmacology in the Egyptian empire equal if not exceed those of modern 4th efflorescence society, and there can be no doubt from the facts of the historical record that such achievements were attained in a spirit of experimentation, as the evidence of fine surgical methods and pharacological results are documented in the exhumation of ancient Egyptian mummies. There is no reason to doubt that the spirit of experimentation that was utilized in these areas of science were not also utilized in the attainment of achievements in other areas, such as physics and engineering. Furthermore, in the fields of medical and pharmacological science, it is undoubtedly true that ancient Egyptian doctors were not limited in their spirit of experimentation by any social, moral or ethical constraints, and it is entirely conceivable that the kinds of experiments they were able to perform on living subjects make even the worst excesses of *in vivo* experiments by doctors in modern totalitarian states pale by comparison.

Given a cumulative amassing of scientific and technical knowledge organized under the political mandate of the Pharaonic clique in the peak period of the Egyptian empire, such a free-ranging spirit of experimentation, where the potential for experimentation was virtually unlimited in the availability of living subjects, makes it understandable how such medical and pharmacological achievements were attained. Under these circumstances there was a constant free-flow interplay between theory and experimentation. Any theory could be tested because experimental subjects were always available from the enslaved population, who viewed the brutalization that such experimentation imposed upon their persons as a religious duty or even a religious joy. In turn, such constant and free-thinking experimentation was always presenting new conceptual ideas to the medical theorists, who could further consolidate the bulk of their medical knowledge by ever more medical experimentation. As in the fields of science and engineering, where energy and labor saving devices are valued not so much for their technical utility as for their social implications in modern society, the same is true for engineering devices adapted to the uses of medical science, such as microscopies, stethescopes, and all other kinds of life-saving devices and apparatus. In such a society as the ancient Egyptian empire, devices such as these in the fields of medical science would have had as limited a utility as devices of mechanical and electro-mechanical nature used in engineering. Except for the lives of members of the Pharaonic clique, and to a much more limited extent certain members of the politically designated technical elite drawn from the Egyptian masses, no life was sacred, and it was not necessary to go to any pains to either save it from death, or ameliorate its pain and anxiety in life.

Thus the variable toxicity of certain medicines or the variable efficacy of surgical techniques were not tested under the controlled conditions of modern science, for the life and health on whom these medical skills and medicines were practiced were of no consequence in terms of the social values of the Egyptian empire. Even with little or no aid from devices similar to modern microscopes and other mechanical devices that test chemical reactions in living metabolism, the range and capacity for trial and error experimentation possible, still a vital component of modern medical research albeit in far more limited circumstances, would have allowed compilation of enormous amounts of medical and pharmacological data. The known efficacy of ancient Egyptian surgery and pharmacology under these circumstances leaves little room for doubt that experimentation in the fields involved included what is today called genetics, since surgery, autopsy, and vivisection, in the context of a vast knowledge of medical chemistry, would have revealed to these ancient medical practitioners the organ sites of hormone production and the chemistry of sexual reproduction. So unconstrained were the circumstances under which these ancient medical practitioners worked that it is possible to imagine that experiments of extraordinary grotesqueness were performed, on both humans and animals, and involving the genetics of both, more extensive and more grotesque than the kinds of genetic experiments involving humans and animals known to have taken place in Nazi concentration camps where designated 'sub-humans' were used in a relatively unconstrained way, by modern standards, as the subjects of such experiments. Thus, the figures which emerge time and time again in ancient Egyptian mythology and among the artifacts of ruins of these civilizations, half human and half beast, such as the Ibis-headed Gods, the bird-headed God, Horace, the human-headed lion of the Sphinx and others may have drawn inspiration from more than imaginary symbols. In all of these cases, however, whatever the consequences in other ways of these experiments, their purpose was always single-minded, the elongation of life on this earth for the Pharaoh and his clique and the reincarnation of the Pharaoh in the same shape and form in a future lifetime.

What findings did result from these experiments and this research was always conceptualized, in part, in terms of subjectification of phenomena and was inextricably mixed with the ancient Egyptian religious system, which was the only way these scientific findings were accessible, or of interest, to the Egyptian masses. In a system of conceptualization that always remained in large part in terms of subjectification of phenomena, any development in one field of cultural or scientific skill always had its counterpart development in every other field of cultural and scientific skill. Religion was that field of cultural and scientific skill which was sanctioned for the Egyptian masses, and since they in fact comprised the vast majority of the population of the empire, it was this field of cultural and scientific skill, in the area of religion and its associated fields of philosophy, literature and mythology, that was best transmitted from generation to generation, and which, like the masses themselves, survived Pharaonic rule even into modern times. The medical and scientific findings of ancient Egyptian research and experimentation may have yielded longer life expectancy for the chosen few and greater freedom from natural diseases among their limited numbers. In addition, the results of these findings have by various devious methods found their way, through Babylonia, and other mid-eastern neighbors of the Egyptians, probably to China, and India as well, and finally to Greek civilization where it was passed on by Hippocrates and others as the basis of modern medical practice and pharmacological and other related biological and biomedical sciences.

The horrific circumstances under which medicine and related sciences had their origins has long been forgotten and indeed undoubtedly deliberately suppressed as the beneficial aspects for a wider range of individuals and collectives in these findings were retained in the convoluted course of the historical process. The major goal for which all of these medical skills were originally intended has been lost however, like the Pharaohs themselves, in the sands of time. The process of human efflorescence reveals the message for present day humanity in the way that it did not for the ancient Pharaohs, that ego awareness is a phenomena that by definition can never be replicated. In each individual, in each generation, ego identity must be born anew, and it is this very special aspect of human efflorescence which so markedly separates human evolution from all prior life forms on this planet. Ego-identity presents humanoids with both the pain and challenge, the hope and despair, which mirror the precariousness of their survival as a continuing life form evolving into the future. Even if all the magic and science of the ancient Egyptians were successful in allowing the rebirth of individuals with precisely the same biochemical makeup that they possessed in a previous lifetime, this would in no way obviate the process of this 'born-again' individual having to once more awaken in pain and in ecstasy to the fact of his own unique individuality under new social and cultural circumstances.

Chapter X

EXPANSION OF THE EGYPTIAN EMPIRE FORGES CIVILIZATION IN THE NEOLITHIC WORLD

From a historical point of view, the concatenation of circumstances that resulted in the Egyptian empire, beginning about 20,000 years ago, are of incalculable consequence. In the Egyptian empire, vestiges of the 2nd, 4th, 3rd and even 1st efflorescence came together in such an explosive mixture that they changed the entire shape of humanoid evolution. Prior to the Egyptian empire it was possible to trace humanoid evolution in terms of collective modes and patterns that markedly differentiated one stage of efflorescence from another. In these prior periods, biochemical and psychic evolution were prime parameters determing the stage of efflorescence, and climatic, ecological, and environmental circumstances were central limiting factors affecting any given stage of efflorescence. From the time of the Egyptian empire onwards, however, the nature of humanoid evolution changed, and the process by which it evolved was largely historical rather than anthropological. While evolution in an anthropological sense involves the external or largely uncontrollable factors mentioned above, such as biochemical, psychological, climatic, and environmental factors, evolution in an historical sense has an entirely different ontology.

The sense of historical evolution, for one thing, implies that mankind no longer evolves or changes solely at the whim of these external and uncontrollable factors. If not entirely the master of his own destiny, he is at least in part responsible for it, in both individual and collective terms. The skills of civilization that had peaked in 3rd efflorescence society among the conquered people of the Nile valley 20,000 years ago, and were organized under the tyrannical reign of conquering Nordic hordes over a period of 15,000 years, were skills that could be transmitted and diffused among all humanoid collectives on the planet. Skills of civilization had attained the status of technology and as such could be rapidly transmitted and diffused even where humanoids aggregated in differing stages of efflorescence, or where these stages of efflorescence had evolved in vague and ill-formed modes. Agricultural science, instituted on a wide scale in selected favorable ecological niches, permitted the genesis and maintenance of large populations who engaged in the fixed routines that the demands of agricultural science necessitated, from generation to generation and from place to place in developed relatively stable and peaceful social orders.

Heterogeneous modes of efflorescence in such circumstances were normalized by the technical and routinized demands that the maintenance of scientific agricultural systems necessitated for the benefit of all individuals in the collective. The science of domesticating animals to enhance the livelihood of large populations led to an institutionalized nomadic way of life which was, at the same time, a necessary part of the learned technology associated with this science and a means by which communication between people separated by large distances developed. As transportation and communication between heterogeneous people separated by large distances became a routinized part of neolithic life in the period 5-10,000 years ago, the partially objectified language systems first developed in ancient Egypt evolved further away from the subjectification of phenomena and towards refined objectification, permitting ease of communication between diverse and separate collectives and enhancing transmission and diffusion of all the skills of civilization.

Where institution of agricultural science was successfully maintained, the skills of city building could be learned and these skills rapidly transmitted to new areas of agricultural institutionalization throughout the neolithic world. Included in the skills of city building were the associated skills and technologies necessary to maintain these cities, such as plumbing, heating, insulation, wall-building, and the partitioning of cities into specialized segments that maintained the welfare of both the individuals of the cities and the collective as a whole. Where civilization extended itself to the shores of the great seas, rivers and lakes, the skills of shipbuilding could be transmitted and diffused, further increasing the capacity for neolithic man to communicate over wide areas. Diversification of livelihood that existed among collectives in somewhat different ecological niches, with the aid of increased means of transportation and communication, permitted the evolution of both trade and warfare as institutionalized means of communication among dispersed collectives in the neolithic world, where both trade and warfare had highly developed associated technologies. Such easily transmittable inventions originating in the ancient Egyptian empire as the wheel, the horse drawn chariot, the smelting of various metals for production of both household goods and weapons, similarly advanced such means of communication as trade and warfare in the neolithic world, and permitted, on the whole, a more sophisticated way of life for all people of the neolithic world, removing them ever further from subjectification of phenomena as the primal mode of conceptualization and psychic engagement.

Those peoples so influenced in neolithic times by the special brand of civilization produced originally in the Egyptian empire as a result of the particular blend of 2nd, middle 3rd, and 4th efflorescence stages of humanoid development included populations extending out from the Nile valley to the modern middle eastern regions, most of North Africa, India, southwest Asia, and Asia Minor, and east to central and eastern Asia. Populations that were also influenced by Egyptian civilization in this way, but remained on the peripheries of extended immersion in this civilization due to geographical barriers, were the populations of the Americas, Polynesia, and northwest Asia. All peoples that came under the direct purview of the Egyptian influence are those peoples today who remain in the late stages of the 3rd efflorescence in the so-called 'underdeveloped' world, or as the modern Chinese have termed it, the 'Third World.' Indigenous Indians of the Americas, northwest Asia, Polynesia, and most of southern and central Africa remained in the early-middle 3rd efflorescence stage of humanoid development for the most part, but where these populations were influenced by less than intermittant contact with the proliferation of skills and technologies from Egyptian civilization, the effect was striking.

Pyramid building technology among certain populations of central and south America, along with other accoutrements in the sphere of religion, science and magic that the Aztec and Inca civilizations possessed suggest direct influence of these populations during the height of the Egyptian empire. The fact that they lacked certain inventions, such as the wheel, and much of the agricultural science and technology base of neolithic civilizations that arose after the height of the Egyptian empire as a result of indirect influence from that Empire testifies to the fact they were likely, at some point, special purpose colonies established by the Egyptian empire during its height, and, as the Egyptian empire declined and these colonies were of no more use to the parent civilization, were left to fend for themselves. Removed as they were by the wide gulf of treacherous seas from the main efflorescence of neolithic civilization which flourished not as a direct result, but more or less in spite of the Egyptian empire, they were unable to make the kind of progress into true late efflorescence civilization that those civilizations contiguous with the Egyptian empire and each other were able to make. The remnants of monolithic construction projects on Easter Island is another case similar to the Inca and Aztec civilizations where, however, the interest in the colonies established there by the Egyptian empire were of such limited interest to the parent civilization that indigenous tribes of Easter Island did not have time to absorb the religious and magical systems which were associated with these constructions, and remained as much in the dark about their origins as the European explorers who happened upon them in the 19th century.

Inca and Aztec civilization still flourished in the 16th-17th centuries, when these civilizations were discovered and conquered and their people enslaved by the Spanish conquistadors, some 12-15,000 years after the peak period of the Egyptain empire and 5-8,000 years after its demise. Archaeological evidence at these sites suggest that these civilizations were in a continuous process of birth and decay, one efflorescence destroyed by internal conflict or external conquest, only to be rebuilt again in time in more or less the same area and in the same way. The continuity of these modes of civilization under such erratic circumstances is testimony to the primal force of the original Egyptian religious, magical and scientific system which seemed to be almost equally comfortable during this period of tens of millenia as when the system was first established in relative geographical isolation from the rest of the neolithic world. After the peak period of Egyptian civilization, when these colonies in central and south America would seem to have been established, the only infusion of new population groups into this area came from wandering collectives of northwest Asia over the Bering straits who apparently eventually worked their way southward, although there were also intermittent infusions of lesser numbers of individuals who came by ship from east Asia to the Pacific coast of north America, and who in time presented threats to these colonies as well. All of these new infusions of collectives to the west of both north and south America, however, did not share in any direct influence from the Egyptian empire, and were also, due to the time period in which they came to the shores of the Americas, not fully established in neolithic civilization as were the populations who existed on a contiguous route from the Egyptian empire.

Neolithic civilization as it extended on a contiguous route from the Egyptian empire beginning about 10,000 years ago spread along this route slowly, so that those areas such as east Asia that were further removed from the neighborhood of the Egyptian empire developed neolithic civilization later than those who were closer. Thus, those groups who sporadically migrated to the shores of the Western Americas from Asia, overland or by sea, were not in possession of neolithic modes of civilization, and as such were usually absorbed into the primal Egyptian modes represented by proto-types of Inca and Aztec civilization whenever they came into close contact with them. In addition, archeological finds and other evidence from east Asia in China from about 5000-7000 years ago indicate that east Asia itself may have originally been the site of an early colony of the ancient Egyptian empire in its peak period. Both Chinese historical records and some early archeological sites in China demonstrate the kind of tyrannical and Pharaonic type control over Chinese civilization at this time.

The tombs of the Shang dynasty about 5000 years ago reveal similar obsessions associated with the death of the king of the country as in ancient Egypt, where all the slaves and concubines of the king were slain on his death, apparently with the purpose of accompanying him on his sojourn through the underworld to eventual reincarnation. Similar kinds of artifacts to ancient Egyptian burial practices are also found in these tombs, models in clay or metal of individual palace guardsmen, their weapons, and their horses. These practices seemed to have been continued at least sporadically in China as late as the Chin dynasty only about 2200-

3000 years ago, as the famous tombs of the Chin emperor attest. The continuation of primal Egyptian modes however by this time in China was extremely eclectic and pejoratively regarded even in that time, as Chinese history and literature documents, since neolithic civilization was well established in China at least 5000 years ago.

In fact the only parts of the world which were untouched by either ancient influence from Egyptian civilization or its aftermath in neolithic civilization, aside from remote areas in central and southern Africa, parts of Polynesia, New Guinea, and Australia, where geographical barriers were so extreme and forbidding, and indigenous ecological features so uninviting that collectives in these areas remained out of contact with the rest of the world since the 2nd and early parts of the 3rd efflorescence, 100,000-500,000 years ago, were the west and northwest of the European continent. In the case of north and northwest Europe it was not that geographical barriers were so extreme or forbidding, or that ecological resources were so uninviting, that either direct or indirect influence from the Egyptian empire was unable to establish itself in these areas. It is also no accident that these areas comprise in modern times the 'developed' world, while the vast part of the rest of the inhabited world, outside of the exceptions where geography and ecology caused those groups to remain in a 'primitive' state until relatively recent times, comprise the 'underdeveloped' world. Even those parts of southern and south-central Europe, such as Greece, the southern part of Italy, and central Europe south of the Urals which were influenced at least in part by the efflorescence of neolithic civilization during and after the demise of the Egyptian empire, remain the least 'developed' parts of the modern 'developed' world.

The reason that what today comprises most of the 'underdeveloped' world was thoroughly influenced by the neolithic revolution in prehistoric times, while north and northwest Europe was not, was that the collectives who inhabited southern regions had either been part of the Egyptian empire during its prime or had been strongly influenced by it, in one way or another, while north and northwest Europe alone had remained out of its purview. The time period during which the Egyptian empire flourished was so long that it may safely be assumed that most areas of the Near East had at one time or another come under its direct influence. In part at least, burgeoning population under the stabilized social rule of the Egyptian empire at its peak was responsible for the spread of collectives carrying modes of Egyptian civilization with them into the further regions of the Near East, southwest Asia, and even further. In part, indigenous populations of areas outside of the Nile valley, while they may not have reached the peak of 3rd efflorescence society, under the influence of migrating, colonizing or invading collectives from the Egyptian empire were eventually assimilated to Egyptian modes, at least to a large extent, when the empire was at its height and very powerful. Nonetheless the social structure of the Egyptian empire was such that any collective out of the direct purview of the Pharaonic clique was of marginal interest only to the empire, and little effort no doubt was made to 'Egyptianize' them in any complete way, their interest to the empire being of peripheral economic or strategic importance only.

In early periods when migrating, or colonizing or invading groups left the center of the empire at the whim of the Pharaonic clique, their strict adherence to modes of Egyptian civilization were no doubt well entrenched. Similarly, collectives who were colonized or enslaved by these migrating Egyptian colonists, in their first flush of attainment in such relatively sophisticated modes, were no doubt deeply attached to them. Over time however neglect by the Pharaonic clique of distant colonies became apparent, since the Pharaonic clique was interested not so much in the collectivity of the empire as is in the individual powers over both life and death such collectivity afforded them. Under these circumstances, the power and influence of the empire understandably began its long decline.

In addition, organization of a political elite among the Egyptian masses, which organization was no doubt spread wherever the empire extended itself, began to produce well-formed vestiges of both 2nd and 4th efflorescence modes among colonized inhabitants. Since the Egyptian empire was not founded on any universal principles of justice or order that extended into the social, political, or economic spheres, and was almost entirely a matter of the deification of individual ego-recognition and libido aggression in collective symbols, disorder and rebellion among distant colonies was inevitable, and was perhaps often inspired by the political elite within the Nile valley as well. On the one hand, rebellion, both inside and outside of the empire, was a result of the political elite among the enslaved masses engaged in a personal power struggle with the Pharaonic clique. On the other hand, it was caused by the ingrained institutionalization of superego among individuals who sprang from the enslaved masses and which alone of the people in these regions, only the Nordic conquerors of the Nile valley had lacked. No doubt in time, and by the time of the full decline of the empire, the descendants of the original Nordic invaders had totally assimilated entrenchment of superego and peak 3rd efflorescence society, just as many of those whom they conquered had absorbed some vestiges of 2nd and 4th efflorescence society that the conquerors had brought with them to that part of the world. As the power of the Pharaonic clique weakened over time, rebellion became a mode of collective behaviour since entrenched superego reacted with hatred and disdain towards the tyrannical excesses of Pharaonic rule.

In the myths, literature and religion of Near Eastern people who inhabited regions adjacent to the Egyptian empire from about 5-8000 years ago, such as the Babylonians and Sumerians, figures who were part human and part animal were still worshipped, but already a human element had appeared. Mankind was created, according to these myths, by cruel Titans, and the major figure of religious worship evolved into the innocuous bull. These cruel Titans may represent the rule of the cruel Pharaohs of Egypt whom the Sumerians and Babylonians won their freedom from with great difficulty, both socially, politically, and psychologically. The religious symbol of fertility which, under Egyptian rule had been the living God that the Pharaoh was meant to represent, was replaced simply by a nameless and docile animal, so that the symbolism of fertility in religion was kept while it was divested of the awful political power it possessed under direct Egyptian rule.

The pejorative way in which female Goddesses and Titans were seen, which was a distinct departure from Egyptian mythology and religious conceptions, reflected the new born ascendancy of individual and collective ego and libido, modes of the 2nd and 4th efflorescence that were new to these regions and had been directly induced by close contact, if not outright submission, in earlier periods, to the Egyptian empire. Of course much of the

Old Testament represents the *summa* of both the historical, religious, and philosophical experience of those Near Eastern collectives whose rebellion against the Egyptian empire and eventual independence from it, had so many vicissitudes. As a collation of this historical and other kinds of experience by near Eastern collectives, the Old Testament reaches its climax with Moses' dramatic exodus from Egypt, the enslaved masses he took with him forming the basis for the truly historically and religiously independent Hebrews in an area remote from the rule of the Pharaohs.

Undoubtedly much of the Old Testament is a collation of myths from various near Eastern collectives who possessed as a common attribute their quest for liberty from Pharaonic rule of the Egyptian empire, and as such these myths are as early as the first rumblings of rebellion against that empire in the Near East, anywhere from 8-10,000 years ago. Rewritten many times in many ancient languages, in its present form it comes down to us in the modern world through relatively modern Hebrew versions from about 5000 years ago and Greek translations first made about the time of Christ. In the English, the King James version is only 400 years old, and these series of relatively modern translations ending with the King James version are attempts to remove these texts, as far as possible, from the modes of subjectification of phenomena in which they were at least in part originally conceptualized. Nonetheless, even in these more modern translations, it is possible to obtain the flavor of the genuine cultural modes of these near Eastern peoples, who veered between subjectification and objectification of phenomena, and whose explications are always on more than one level at the same time.

Beneath the historical exposition of the quest for liberty, there always lies the foundation of the organized fields of knowledge developed under the aegis of Egyptian civilization which played such an important role in the extension and diffusion of neolithic civilization. This foundation remains the basic context in which all of the religious philosophies of the 'underdeveloped' world evolved to their present form, including the religious philosophy propounded by the Old Testament itself. In the text of the Old Testament, science, history, religion and philosophy are all inextricably mixed, although there is in the Old Testament, as in other texts that form the basis for modern day religious philosophies of the 'underdeveloped' world such as the Koran in the Near East, the Upanishads of India, the texts of Buddhism of India and China, the texts of Confucianism, Mencius, and the Tao in China, the tendency to underplay the importance of science as such to humankind. Science as one of the earliest developments in human evolution is directly associated with the evolution of ego-recognition and libido aggression, and in the hands of the Pharaohs of Egypt who deified their own ego and libido in terms of the collective it became a weapon of extraordinary cruelty and oppression for the masses who suffered at its hands.

In all of the religious philosophies of the 'underdeveloped' world which are ego-negating and libido abdicating, there is an almost involuntary anti-science bias. In addition, in these philosophies science is often directly associated with magic, where both science and magic are seen as contrary to the general ego-negating and libido abdicating thrust of these philosophies, detrimental to both the individual and the collective as a whole, and contrary to the will of both cosmic consciousness and cosmic libido, whose purview is always seen as clear and distinct from that of man's. So traumatic were the experiences of those collectives who suffered under the reign of the Egyptian Pharaohs that, like all traumatic experiences, the specific means by which trauma was produced was psychologically suppressed. The trauma of these collectives was both individual and collective, it lasted for such a long time and was so thorough and extensive, that the term 'oppression' suffices to cover the multitude of ways, the length of time, and the thoroughness with which this trauma was induced. It yielded a collective neurosis, as real and debilitating as an individual neurosis, and with the same kind of reluctance or even inability to examine the modes of its original genesis. Like individual neurosis, which most people suffer from in various degrees, however, collective neurosis can evolve a pattern of life-adjustment and may yield new modes of extraordinary inventiveness and adaptive advantages. While people in the 'developed' world often regard those from the 'underdeveloped' world as innocent or childish, timorous and even cowardly in not aggressively attaining those modes of individual and collective behaviour that will allow them full and proper entry into the 'developed' world, people from the 'underdeveloped' world seem afraid of using modes of force and conflict, struggle and competition, which ultimately yield such tangible, material advantages, both individually and collectively.

But, from the point of view of the 'underdeveloped' world, people of the 'developed' world are like children who dare to venture into an unknown forest, convinced that no tigers or other vicious animals of prey lurk there because they have never seen them before, while people of the 'underdeveloped' world are like old and craven men who refuse to venture in these wilds, because they have been there in their youth and know exactly what horrors await them for their pains. Thus the people of the 'underdeveloped' world would prefer to remain in the late stages of the 3rd efflorescence rather than to rush headlong into the 4th, since, because of the sobering legacy of their direct experience with the Egyptian empire thousands of years ago, they have instinctive, if not always rational, fears of the inevitable consequences of social and individual misery that can result from a combination of aggressive modes of 2nd and 4th efflorescence society.

Those people of the 'developed' world in north and northwest Europe and their derivative societies in the Americas, Autralasia, and 'white' South Africa have neither the collective conscious or collective unconsciousness which would lead them to hesitate before leaping into 4th efflorescence society. What vestiges they have of the sobering legacy of direct experience with the Egyptian empire comes to them in bits and pieces, mostly through the rationalized version of the Hebraic Old Testament and the religious philosophy of Christianity, which is a combination of this Hebraic tradition and the influence of Greek philosophy, a philosophy almost entirely untouched by any direct contact with the influence of the Egyptian empire. In addition, the Old Testament, even in its most rationalized English version translated in the time of King James, presents the subject of the experience of the Near Eastern people in revolt from the Egyptian empire in a manner that is in great part still conceptualized in terms of subjectification of phenomena. As such, much of both the emotional and cognitive effect and intent of this text remains conceptually inaccessible to people of the 'developed' world who are accustomed to conceptualizing only in terms of the objectification of phenomena.

Even though collectives of the 'developed' world have institutionalized superegos to greater or lesser degrees, depending on their particular collective history, none have had the experience with the semi-subjectified, semi-objectified modes of conceptualization that were the particular legacy of the Egyptian empire. Even in the cases where superego was most thoroughly institutionalized among the antecedents of collectives in the developed world, from about central France southwards to Greece, evolutionary development was such that these collectives passed through the peak of the 3rd efflorescence and jumped immediately into the 4th where objectification of phenomena was the mode of conceptualization. There was no period for these collectives when the fields of knowledge attained in the peak of 3rd efflorescence society were organized in semi-subjectified, semi-objectified modes of conceptualization. Their entry into the 4th efflorescence was impelled by the acquisition of objectified modes of conceptualization that they obtained from the tail end of neolithic civilization in the Mediterranean and Near Eastern areas, never having undergone the experiences which ultimately led to these end points of neolithic civilization in these areas.

The Greek civilization was founded by the roaming Nordic hordes who lived by piracy and banditry near the end of the neolithic period of civilization about 4000 years ago. These particular Nordic Greek hordes however had already undergone peak 3rd efflorescence evolution and were not nearly as aggressive as the Nordic hordes that had originally conquered the people of the Nile valley some 20,000 years ago. At the point at which they were exposed to neolithic civilization in the Mediterranean area, those civilizations had themselves reached a peak point in objectification of phenomena in both language and thought, as a result of the experience of the wide dissemination and extension of neolithic civilization that had occurred in those areas over many thousands of years. Thus the basis of Greek language was acquired from the Phoenicians who already had a phonetic alphabet by the time the Greeks came into contact with them some 3-4000 years ago. As the quickest and most efficient way of extending the trade and warfare necessary to the dissemination and extension of Neolithic civilization at this time, particularly to remote collectives and areas that had previously not been in contact with Neolithic civilization, the Phoenicians, an aggressively sea-faring offshoot of Near Eastern civilization, had developed a phonetic alphabet. While the phonetic alphabet remained for the Phoenicians only as a form of short-hand, or a code, which allowed them to abbreviate, for purposes of expediency, the wealth of knowledge contained in their modes of semi-subjectification, semi-objectification of phenomena, it was eagerly adopted by the Greeks entirely out of this context. It was extended and elaborated by the Greeks, who had previously possessed little if any rudimentary forms of written language, and became the basis of a language system that was almost entirely an expression of objectification of phenomena.

The wandering of the Greeks, as relatively benign strangers in the neolithic world near the tail end of its present form of evolution, brought them again and again face to face with the strange and marvellous phenomena that was the ultimate legacy of the Egyptian empire, and which already at that time was no longer recognizable in its original form. The phonetic language they had adopted from the Phoenicians and which enabled them to formulate and crystallize the objectification of phenomena as a mode of thought in written language was used by the Greeks to record, in an objectified form, all sorts of phenomena which were inaccessible to them by virtue of their lack of experience in both the modes and collective history in which these phenomena were shaped. The documentation of the early experience of their collective under these circumstances became the legacy of Greek civilization in terms of their myths and religion which they bequeathed to the antecedents of the modern 'developed' world. It was also the basis on which Greek civilization as such was founded, with its objectification of phenomena in all fields of knowledge, and which became the legacy of civilization upon which that of the 'developed' world in the 4th efflorescence mode was based. These fields of knowledge, like the myths and religions of the Greeks, which were both conceptualized and documented in terms of the objectification of knowledge in a phonetic written language, were also eclectically acquired from the tail end of the neolithic civilizations from 2-4000 years ago. In this way the Greeks were the ultimate heirs to the accomplishments of the Egyptian empire in all fields of knowledge and skills of civilization.

The almost total objectification of phenomena in these fields of knowledge and skills of civilization, which is the legacy of the Greeks in modern 4th efflorescence society, deleted those parts of these fields of knowledge and skills of civilization that were conceptualized in terms of the subjectification of phenomena. On the positive side of this deletion, it allowed full rational accessibility to all fields of knowledge and skills of civilization that they had acquired by cultural diffusion from middle eastern civilizations heir to the legacy of the Egyptian empire. On the negative side of this deletion, however, the actual historical process by which these knowledge and skills had been acquired and the emotional and cognitive affects that resulted from this process were lost to Western civilization. Such deletion of course was not deliberate, nor was it the result, as in the Hebrew and other later mid-eastern and Asian versions of collective history, of a religious or moral imperative of any kind. The deletions resulted simply from the fact that the Greeks had no way of conceptualizing in terms of the subjectification of phenomena, not at least in any organized way, and the kind of written language they adopted from the Phoenicians allowed them to perfect conceptualization in terms of objectification of phenomena.

While the Greeks and their antecedents had undoubtedly passed through 3rd efflorescence society and had a well-established institutionalization of superego, their accomplishments in the 3rd efflorescence paled before the organized modes of 3rd efflorescence society they encountered in their wanderings among the heirs to the Egyptian empire. Ready as they were, from an evolutionary point of view, to enter into 4th efflorescence society, they left behind them their comparatively meagre accomplishments in 3rd efflorescence society for the riches they could cull, through cultural transmission, from Mediterranean civilization for the purposes of establishing themselves in a 4th efflorescence mode. Only in the most modern times, where European and especially German civilization in its more positive aspects in the 19th and 20th century, has allowed further development in conceptualization in terms of the objectification of phenomena, does the legacy of the Greeks in all of its ramifications become accessible in terms of rational analysis. The work of such great modern German theorists in this mode such as Freud, Jung, Marx, Simmel, and Weber, and French theorists such as Durkheim, Sartre and Le'vi-Strauss, and English theorists such as Darwin, allow us finally to conceptualize in terms of the objectification of phenomena those aspects of the legacy of the modern world, both in the 4th efflorescence and the late stages of the 3rd, that belong truly to the purview of conceptualization in terms of subjectification of phenomena.

Freud developed the theory of psychic dynamics and illustrated the various devices by which, in personality malformation, parts of the psychological life-history of the neurotic or psychotic individual are suppressed and often transformed, through displacement, into entirely fictitious events. Jung extended Freudian theory to the collective material of myths, in which he was able to pinpoint these same developmental processes and self-protective mechanisms that stretched, not over the lifetime of an individual, but over the millenia that marked the lifetime of an entire collective or race. Le'vi-Strauss, the great modern French anthropological theorist, extended these theories even further in his studies on myth. The structural components of myth, he posed, were not merely a reflection of collective psycho-dynamics but also expressed absolute modes of thought common to all humans in all times and places. In addition these structural components of myth also reflected both the social organization and social experience of a given collective where particular details of that social organization and experience, although extant in the myths as a structural component, were unknown due to loss of information about them over time.

Although most of Le'vi-Strauss' work was with the Indians of South America, if we extend this general approach to myths of civilization in the late stages of 3rd efflorescence society, such as the myths of the Babylonians and others in this region and their culmination in the Hebraic Old Testament, many of the structural components of these myths which constantly appear in various forms may be seen from the point of view of the political oppression which these groups suffered under the ancient Egyptian empire. The obscurity of some structural components of myths in the Old Testament which also reflect similar myths in Babylonian and other Near Eastern cultures, and even Asian cultures, is a result of the psychological mechanism of repression. In these myths repression as a result of psychological experience too painful to bring to rational consciousness is closely associated with deliberate supres-sion of facts and events that offended the moral and political sensibilities of the writers of these myths.

Chapter XI

ESCAPE FROM EDEN: THE HEBREWS ESTABLISH A CIVILIZATION IN OPPOSITION TO THE EGYPTIAN EMPIRE

Thus, in the Book of Genesis, for example, as well as in other parts of the Old Testament, although there is an attempt to organize material of the text in terms of the objectification of phenomena in line with the actual history of the text itself, the material expressed is meant to be understood on many different levels at the same time. While the figure of God in the Book of Genesis is no doubt a valid religious expression of the cosmic consciousness, the personalization of this total control over the minds and bodies of Adam and Eve suggest both a repressed and suppressed historical series of figures who were the oppressors of those original collectives of the Near Eastern region which Adam and Eve are meant to symbolize. The actual genesis of the natural world, and its biological evolution leading to humanoid development, are precise scientific statements of natural and biological evolution which rival Darwin's in their precision. This aspect of the myth of Genesis illustrates again the powerful scientific capacities of early humanoid collectives. The close relationship in time and morphology between the evolution of animals and man, in specified genuses and species, was necessarily passed by oral tradition through the 2nd and 3rd efflorescence of humanoid development to the later period in which this text was actually compiled in written form.

In accurate biological order, the Book of Genesis lists the efflorescence of the plant kingdom as prior to that of the animal kingdom, and of the animal kingdom, creatures that flourished in the seas were the first to appear. Interspersing modern discoveries, these creatures included the bacteria, fungus, insect, and Crustacea which were the first members of the animal kingdom to appear in biological time. Next came the fish and the fowl, with whom reptiles of the sea and amphibious reptiles were also apparently and correctly included. Finally mammalian evolution took place on land, concurrently with further reptilian development as suggested by the 'creeping things' that flourished along with mammals. In the last, man evolved.

Much of the text as well reflects the facts of man's psychic, and psychogiological evolution. The early chaos of the natural world before God imposed order on it reflects the subjective experience of man in his 1st efflorescence as seen from the perspective of later stages of evolution, where man was without 'light,' without knowledge, or any sense of order in either his external or internal world. Of prime importance from the perspective of man's psychic and psycho-biological evolution is the fact that Adam was created before Eve, reflecting the insight that in his earlier stages of evolution only the 'male type' was psychically well-formed. The absence of a well-formed 'female type' is characteristic of humanoid evolution in its second efflorescence. Thus, when Adam existed alone prior to Eve he had all the consciousness and knowledge that he possessed after the creation of Eve, even in this period possessing the faculty of language which enabled him to give names to all the various living species. This reflected the insight, and perhaps even a clear memory, that human society can and did exist without a well-formed 'female type.' That Adam and Eve represent the psychic, rather than biological evolution of man, is seen from the fact that in a passage prior to the genesis of Adam and Eve as distinct human personalities, the Book of Genesis notes that both males and females of every species had been created by God, including the human species. The creation of Eve by God is ascribed to His attribute of mercy, suggesting at one and the same time the nature of the cosmic mind and also that attribute of mercy which the development of the 'female type' in 3rd efflorescence society brought to human evolution: 'It is not good that man should be alone; I will make him a help meet for him.'

The fact that 'Woman' was made of a rib removed from 'Man' reflects an insight of a deep biological, and biochemical nature, and since much of this text was compiled after the height of the Egyptian empire, it is even possible that it reflects biological and biochemical knowledge of an extremely precise nature. Hormonal systemics that differentiate male from female differ not in substance but in minute structural variations. In modern laboratories it is a simple matter to transform male hormones to female hormones and vice-versa, and this was no doubt akin to the possibilities in the medical laboratories of the ancient Egyptian empire. From a psycho-biological point of view as well the 'male type' is associated with the lower nervous system possessed by reptilian species and others prior to mammalian development, while the 'female type' is associated with the upper nervous system that evolved only in mammalian species, finding its final psycho-biological form in the 'female type' of humanoid evolution. These facts, in concert with the statement that God made Adam 'fall into a deep sleep' so that he could remove his rib to make the 'female type,' Eve, and which appears to refer to both exquisitely sophisticated surgical and anaesthetic procedures, suggests a wealth of knowledge about biochemistry, biology and medicine that reflects the direct experience of these Near Eastern groups with the highest achievements of Egyptian civilization.

When, in a later chapter of the Old Testament, Exodus, an apparently illegitimate prince of the Egyptian empire, Moses, is elevated as an important leader of the Hebrew race, to whom the authorship of the Old Testament is attributed, there is a strong indication of just how closely this Near Eastern collective was associated with the technical and political elites of the ancient Egyptian empire. Even in Genesis, before the advent of Moses onto the historical scene, one of the later descendents of Adam, Joseph, a younger son of Jacob, achieves by a process of convoluted historical vicissitudes, the post of chief minister to the Egyptian Pharaoh of his time, again reflecting the intimate association of the Hebrews with the highest technical achievements of the Egyptian empire.

In the garden of Eden, which is east of the eventual areas in which the Hebrews settled, as is Egypt, Adam and Eve live in a state of innocence and bliss, in which all of their material needs are attended to by God. This structural component of the myth may be read at least on two levels at once. On the one hand, it reflects the relative bliss and innocence which humans enjoyed in the 3rd efflorescence of their development, when the individual ego-recognition and libido aggression of the 2nd efflorescence became dormant by virtue of the institutionalization of superego. In this 3rd efflorescence state humanoids felt totally at one with the natural world, both external and internal, viewing as they did the infusion of the moral force emanating from their own superegos into all phenomena. In this state of efflorescence, all conceptualization was in the form of subjectification of phenomena, and there was little or no sense of conflict in either the natural or human world. Seen from the point of view of psychological transformation, however, which occurs in myths, both individual and collective, it reflects the obverse of the actual emotional affect. That which was unpleasant was transformed to something pleasant, where both psychological mechanisms of transference and displacement were operative.

The lack of consciousness and identity of Adam and Eve here reflects the actual suppression of identity and consciousness of the Egyptian masses by the political oppressors, who were the Egyptian ruling elite. All the material needs of Adam and Eve were attended to, but in a way that masters attend to the needs of their slaves. So total was the enslavement of the Egyptian masses represented by Adam and Eve that they were denied even a sense of their own individual sexual identities. The obverse suggestion that in the 'Garden of Eden' the Egyptian masses were no better than herds of domesticated animals, tended by their masters and used at their whim, suggests that, at least to some extent, the experience of the authors of this work involved collective modes of the 1st efflorescence, where collectives still in this mode were captured, possibly from central Africa, and brought to Egypt where they were enslaved. This possibility is reinforced by the total sense of innocence Adam and Eve enjoyed in the Garden of Eden, an innocence so complete it suggests a direct leap from 1st to 3rd efflorescence modes without any direct intervening experience of the 2nd.

Having been brought to the Egyptian empire still in a 1st efflorescence mode, the culture they were absorbed into was 3rd efflorescence, and the total lack of guile which Adam and Eve display suggests a dawning of consciousness under these kinds of circumstances. If this were the case, as it appears to be suggested, the overwhelming attraction of the tree of knowledge for Adam and Eve, which represents ego-identity, is all the more understandable. Having attained 3rd efflorescence modes without ever having undergone 2nd efflorescence modes, and with the guileless approach, in spite of all pejorative warnings to the contrary, that ego-recognition represented, the lure of the 'tree of knowledge' proved irresistible. Of course, the model of 1st-3rd efflorescence collectives is an ideal model, even from a close reading of the text itself, and suggests only that this kind of experience was merely part of the collective experience of those groups who lived under the direct influence of ancient Egyptian rule.

The collective experience symbolized by Adam and Eve is compiled, fused and compressed and no doubt represented varieties of collectives with vestiges of modes in various combinations. The pure innocence and lack of guile which the authors wished Adam and Eve to symbolize however were best expressed in the kind of emotional affect that resulted from 1st-3rd efflorescence combinations as an ideal model, both best reflecting the moral purity of the enslaved group as opposed to the enslavers, and also as best reflecting, from an evolutionary point of view, the experience of pure 3rd efflorescence society where 2nd efflorescence modes had been suitably and effectively tamed.

Adam and Eve were forbidden to eat of either the tree of knowledge or the tree of life, where the former symbolizes ego-recognition and the latter symbolizes libido aggression. After Adam and Eve eat of the tree of knowledge, they are expelled from the Garden of Eden, partly as punishment for their sins in disobeying God's commandments, and partly because God is afraid that if they are allowed to stay in the Garden of Eden any longer they will also eat of the tree of life, and become immortal like Gods themselves. While the tree of knowledge represents ego-recognition, the tree of life represents libido aggression. From the moral and religious point of view the tree of knowledge represents cosmic consciousness which man is, under severe penalty, allowed to attain, but cosmic libido aggression and identification with it is disallowed under any circumstances. This is one of the precepts of religious philosophy where the Hebraic religion differs from those of the Far East. In this sense, there is a similarity of Hebraic with Greek philosophy in its mode of objectification of phenomena where a distinct separation is made between human life and the universe itself.

From an historical point of view, however, the tree of life, or cosmic libido, has a strong moral pejorative, since cosmic libido or the God-like powers it confers on man represents the oppressive powers which the Pharaonic elite exercised over the Egyptian masses and other near eastern peoples under their direct control. The prohibition against partaking in cosmic libido and the consequent powers it endows over the workings of man and nature had the ultimate effect of producing a powerful anti-science bias in this form of religious philosophy. Both the tree of knowledge and the tree of life have direct associations with the pharmacological and medical technologies of the ancient Egyptians, referring as they do to special powers and abilities received as a result of both understanding and consuming specific herbal or medicinal substances. While the fruit of the tree of knowledge merely allows man to extend his powers of both ego-recognition and rational and cosmic consciousness, the tree of life, however, allows man to extend his influence over the natural universe and all living creatures in it by changing his actual biochemical systemics to permit extension of individual life and giving him God-like powers.

This premonition of immortality inherent in the experience with the tree of life suggests the Pharaonic obsession with immortal life through reincarnation which was meant to be achieved through the utilization and application of science, and which, from the point of view of these people enslaved by the Pharaohs, suggested the purposes for which their political and social oppression was so severe and unbearable. As a result of their immediate historical experience, therefore, science as such was faulted as basically immoral in and of itself, leading as it did to ultimate oppression of a political, social, psychological and even religious nature. If man were to obtain ego-recognition and both rational and cosmic consciousness, and direct individual relationship with God, it was necessary to forego science and prohibit it as basically immoral. We see here then the extension of ego-recognition and both rational and cosmic consciousness from an individual mode to a collective one, since it was science in the hands of the Pharaohs which prevented ego-recognition not only on an individual basis alone among the enslaved population, but also collective ego-recognition.

Perceived by their masters as little better than a domesticated herd of cattle, not only was individual ego-recognition impossible but even more to the point, any sense of collective ego-recognition was impossible. Thus, when Adam and Eve eat of the tree of knowledge, they not only gain individual ego-recognition, but they also gain collective ego-recognition. Each senses their own identity and at the same time the identity of the other: 'And the eyes of them both were opened, and they knew that they were naked; and they sewed fig leaves together; and made themselves aprons.' From this collective ego-recognition, procreation of the race is possible, and all the generations of Adam are created. The making of aprons or clothes by Adam and Eve after having eaten of the fruit signifies that the two of them together, after having achieved individual ego-recognition, also achieve collective ego-recognition. The hiatus of logic which first claims that all species, animals and humans, after their creation were able to reproduce themselves after their own kind, and the fact that Adam and Eve did not produce offspring until after they had eaten of the fruit of the tree of knowledge and were expelled from Eden, illustrates the symbolism of Adam and Eve having attained not merely individual ego-recognition in eating of the tree of knowledge, but also collective ego-recognition. Thus the true human group, i.e., a procreating group, appears only after collective ego-recognition has been attained.

The serpent who persuades Eve to disobey God's wishes and eat of the tree of knowledge after which she persuades Adam to do the same, echoes the Egyptian myth where the 'female type,' the Goddess Isis, uses the serpent to force Ra to reveal his true identity to her. The nuances of the Egyptian myth and this aspect of the Hebraic one are however significant in that the Hebraic one continues to see the problem of ego-recognition from both a moral point of view and from the point of view of the enslaved peoples of Egypt rather than the enslavers. In the Egyptian myth, the serpent as a surrogate of the 'male type' assists Isis in attaining individual identity as the 'female type' in contradistinction to the 'male type.' In addition, since Ra has such a reluctance to acknowledge his own individual ego-identity, we can assume that it is because of his proclivity to identify himself with cosmic and collective forces, rather than acknowledge his own psychic drives as individual and hence terminable. Isis' role in the Egyptian myth is a humanizing one, where at least within the ruling clique of God-Kings and Queens, there is an attempt to establish human 'male types' and 'female types.' This humanization in the Egyptian myth is clearly confined to the ruling elite itself, as both Isis and Ra never lose their general status as Gods or God-Kings and Queens.

In the Hebraic myth however the expressed intent of the surrogate 'male type' represented by the phallic serpent is to make both Adam and Eve God-like through eating of the tree of knowledge. However, once they have complied with the serpent's wishes, in defiance of God's command, the opposite effect is achieved from the Egyptian form of the myth. By attaining ego-recognition, both individual and collective, they lose their lien on immortality and are thrust out of the garden of Eden where they will work by the sweat of their brows and return to the dust of which God tells them they are made. Eve is singled out for particularly severe punishment where her sin, as the one who inspired Adam to disobey God, keeps her enthralled to her husband, in status somewhat like that of servant or slave. General thematic components which run through the Old Testament makes these subtle nuances of difference between this myth and the Egyptian one clear, especially when seen in the light of the existing social and historical circumstances of the Near Eastern races it describes.

Ego-recognition, as directly associated with the 'male type' rather than the 'female type,' is a mode of 2nd efflorescence society in the general trend of evolutionary development; in the case of the enslaved peoples of the Egyptian empire however, who were either in the 3rd efflorescence state, or had leaped directly from 1st to 3rd efflorescence state under the conditions of their enslavement, there were few vestiges of 2nd efflorescence development. What vestiges existed were deliberately suppressed by the ruling Egyptian elite as possession of these modes made the enslaved groups a political and social threat to their own power. Without possession of at least some 2nd efflorescence modes, no group of the 3rd efflorescence could enter into a later phase, and for this reason we see Adam and Eve, after having eaten of the tree of knowledge, gaining a sense of both individual and collective ego-recognition almost simultaneously.

Thus, in order for enslaved groups of the Egyptian empire to establish themselves as politically independent from the rule of the Egyptian empire, it was necessary to reacquire the modes of the 2nd efflorescence which they had lost, or in some cases, never possessed, in order to obtain collective ego-identity and political liberty. Thus in the Hebraic religious philosophy, which remains mostly of an ego-negating and libido-abdicating philosophy, for historical and political reasons, modes of the 2nd efflorescence are idealized and even to some extent deified, in order that the collectives involved are able to establish themselves as politically and socially independent of Egyptian rule. From an overall point of view, the ends are used to justify the means, where the ends are the religious philosophy of ego-negation and libido-abdication, and the means are the induction of modes from the 2nd efflorescence which ultimately result in a collective sensibility that allows the establishment of the morally required religious and philosophical system. In this way, Eve's ego-identity in terms of the 'female type' can be sacrificed, because in fact the entrenchment of institutionalized superego in these collectives is strong enough to permit such a sacrificial gesture, a sacrificial gesture which is in fact nominal rather than substantial under the circumstances of well-entrenched institutionalized superego. While in the Egyptian myth the serpent as surrogate of individual male ego-recognition and libido aggression remains both idealized and deified, in the Hebraic myth it is punished and cursed.

The enmity that God produces between the serpent and the woman, in particular, is another reference to the submission that the 'female type' must endure under the 'male type' where 2nd efflorescence modes are to be revived, for the sake of both the collective and the ultimate establishment of the Hebraic religious philosophy. The fact that a general enmity between the race of serpents and all human beings is produced however, where the woman's seed 'shall bruise thy head, and thou shalt bruise his heel' is a reference to the moral prohibition against the acquisition of libido aggression by male and female alike. Even though 2nd efflorescence modes have to be reacquired for the achievement of collective and religious ends, they are never meant to be allowed to become an end in themselves. Libido aggression is allowed within the range of certain limited circumstances, particularly in the relationship between male and female, but it is not permissible to exceed these limitations, for reasons that have their roots both in the religious philosophy of the Hebrews and the social and political circumstances from which they are derived. The enmity of the serpent towards man echoes the threat of the tree of life which is ultimately forbidden by God to man. Libido aggression is permitted, only within the limits prescribed by God; all other manifestations of libido, on individual and collective levels, and identification with it on a cosmic level, is prohibited and must be fiercely resisted by the self-conscious individual who is directly attuned to the cosmic mind of God.

Of course it could be argued that the inhibition about expressing libido aggression results from a combination of guilt and fear induced in the enslaved masses of the Egyptian empire by the Pharaonic elite in terms of being allowed to express their own libido aggression contrary to the will of this elite. This argument, however, while perhaps containing a grain of the whole truth would overlook the powerful moral and spiritual repulsion which these enslaved peoples felt towards the overwhelming libido aggression of their masters, for which aggression they were so often the helpless and unwitting subjects. It was in fact the reactive repulsion felt by the enslaved masses to this overwhelming ego-recognition and libido aggression of the Pharaohs that inspired the total religious philosophy of the Hebrews. In this context, the pejorative aspects of even tasting of the tree of knowledge, which represents ego-recognition, can be understood. While the Hebrews recognized that ego-recognition was necessary if they were ever to free themselves from the tyrannical will of the Egyptian empire, they still regarded it with a great deal of moral distaste because of their own direct experience with its consequences in the hands of untamed leaders.

It is this general distaste even for ego-recognition as such, as well as libido aggression, that orients the Hebrew religious philosophy towards ego-negation and libido abdication, placing it fully in the context of late 3rd efflorescence development rather than in the context of 4th efflorescence proper, however there might be premonitions here and there of 4th efflorescence society. The vestiges of 4th efflorescence that do exist and are encouraged in the Old Testament, like vestiges of the 2nd, are always regarded as unsavory necessities, rather than ends in themselves. Thus, the idealization and even semi-deification of the 'male type' that persists throughout the Old Testament reflects not valuation of ego-recognition and libido aggression that are traits associated with the evolution of the 'male type' in the 2nd efflorescence, but rather the rational consciousness of humans that develops in the late stages of the 3rd efflorescence or in the 4th, and through which a direct relationship with the cosmic consciousness or the mind of God is effected. The tendency for the 'female type' to be devaluated in the Old Testament is more apparent than real, and results from the social and political circumstances that demand some concerted attempt to revive those traits of the 2nd efflorescence and propound those traits of the 4th through which rational and ultimately cosmic consciousness can be attained.

The continuing attempt to personalize God, or the cosmic mind itself, in terms of a 'male type' echoes the confluence of seemingly contradictory themes which are the expression of more than one set of phenomena, religious, social, political and psychic, represented through the same sets of symbols and through a single expository narrative. This expression of many differing levels of phenomena and thought in a single set of symbols and expository narrative is common to most of the religious philosophies that remain in the late stages of the 3rd efflorescence, and makes these philosophies in many ways self-limiting. Since the religious philosophy they express is so closely bound up with the particular history and experience of a certain collective group, it is difficult for these religious philosophies to have universal appeal, or even to be understandable to people who do not derive from the given collective and share the assumed values and attitudes of the group as a matter of collective conscious and unconscious both.

The inability of the Old Testament as a religious philosophy to be universally appealing is self-consciously realized in the text itself, as there is an increasing emphasis as the narrative continues on the distinction made between the Jews, who are the 'chosen people' and the rest of the world. This emphasis on shared racial and cultural values is a recognition of the fact that for individuals outside of the collective as it evolved in time from its roots in the ancient Egyptian empire, the ability to achieve the ends of ego-negation and libido abdication will be extremely hampered, based as these ends are on the convoluted means of social and psychological adjustment of a particular collective in particular historical circumstances. The way in which God continues to be personalized in terms of a 'male type' is a case in point. He communicates with man directly through a medium resembling spoken language. This is a reference undoubtedly to the semi-subjectification, semi-objectification of phenomena modes that were expressed in both written and spoken language in the ancient Egyptian empire. As the Near Eastern peoples who revolted against the Egyptian empire became truly independent of it, their language forms resembled more the objectification of phenomena modes than the subjectification ones. Yet the kind of political and social power associated with this original mode of communication remained a clear memory for these now independent Hebrews, and invariably the libido aggression of a personalized God was more easily appreciated by these peoples when they portrayed his communication in the same form as their distant masters had once communicated with them when they were slaves under the rule of the Egyptian Pharoahs.

As the narrative of the Old Testament unwinds, there are continuing concerted attempts to separate, according to differentiated social and collective modes, those groups who truly belong to the 'chosen people' and who will therefore be able to absorb the otherwise convoluted steps by which ego-negation and libido abdication may be achieved, and those who will not. The conflict between the first two sons of Adam and Eve, Cain and Abel, illustrates the first of many cases where a group with social and collective modes different from the 'chosen people' are proscribed in terms of their opportunity for achieving the moral ends of the philosophy. Again, the set of symbols used and the narrative that is expostulated expresses, in a semi-objectification mode, more than one phenomenological level at the same time. From the point of view of evolutionary development, the struggle between Cain and Abel and their fierce competitiveness for the favour of God, who on one level is a father surrogate, represents the conditions of 2nd efflorescence society where the 'male type' is well-established and the 'female type' is not. The ultimate punishment for Cain in his neolithic aggressiveness expressed in the slaying of his own brother, Abel, represents the moral pejorative against 2nd efflorescence modes which exists as a major, although disguised, theme throughout the Old Testament.

On another level, the differentiation between the good brother Abel, who is the herder of cattle, and the evil brother, Cain, who is a farmer, represents the social limits of acceptability in the collectivity of the 'chosen people.' As a farmer, Cain resembles far too closely the social and economic mode of the enslaved masses of Egypt, from whom the 'chosen' people attempted to differentiate themselves in order to become politically and socially, and also economically, independent of their Pharaonic masters. Mode of livelihood here is made a criteria of acceptability in the collective of the 'chosen people.' These themes are repeated again in the later conflict between Jacob and Esau for the favour of their father, Isaac, and the ultimate inheritance of his patrimony. Again, from the point of view of evolutionary history the representation of the ultimately disfavoured son, Esau, as a hunter, and Jacob, the ultimately favoured son, as a herder of cattle and sheep, represents the conflict between modes of the 2nd and 3rd efflorescence. Although Jacob does not slay his brother Esau to gain his father's favour, with the willful and conscious aid of his mother, he succeeds in deceiving his father and outsmarting his brother in order to win his father's blessing, which seals his inheritance and results in the disenfranchisement of his brother.

Jacob's deceit is not viewed in a pejorative way, since his brother's willingness to sell his birthright for a bowel of porridge reveals the inferior and more primitive modes of 2nd efflorescence society as opposed to 3rd, represented by Jacob's guile in the interests of fulfilling what he believes to be his moral, rather than natural, rights, in the inheritance and stewardship of his father's patrimony. The fact that Jacob performed his acts of deceit with the aid of his mother who symbolizes institutionalized superego, further reinforces the representation of Jacob as the model of 3rd efflorescence modes. Jacob and Esau are further differentiated by the fact that Esau is portrayed as a 'hairy man' while Jacob is hairless, suggesting again the more primitive and hence inferior nature of 2nd efflorescence modes in comparison with 3rd. The fact that Jacob's father Isaac in fact prefers his hairy first son, Esau, the hunter and has to be deceived into giving his blessing and hence his patrimony to the less-favoured Jacob represents the denigration of the 'male type' as personified by the father. In this scenario the father, Isaac, may also represent the Pharaonic elite which favours 2nd efflorescence modes, but which can ultimately be manipulated by the technological, religious, and magical knowledge of 3rd efflorescence modes, so that Jacob, on this level, represents the politically designated elite of the enslaved masses of the Egyptian empire.

When Jacob has to flee into the desert to escape the wrath of his father and brother after having deceived them both and winning his father's blessing, he is forced to engage in a struggle with the angel of death which he wins. The struggle between Jacob and the angel of death represents at one and the same time the libido of 2nd efflorescence modes in conflict with the superego of 3rd efflorescence modes, and also the disadvantage of the Pharaonic elite in employing technological, magical and religious knowledge against the rebellious slave elite, which is after all destined to succeed because the slave elite possesses more of this knowledge and more precisely than the Pharaonic elite. On this level as well the angel of death represents cosmic libido which is identified with the Pharaonic elite, while Jacob, as a morally superior being, has cosmic consciousness, or the mind of God on his side, and therefore ultimately triumphs. In this particular myth or allegory, we see again where the religious philosophy of the Hebrews and Near Eastern peoples in general differs from the religious philosophy of those peoples in the late stages of the 3rd efflorescence farther to the East. Cosmic libido, identified so closely with Pharaonic power, as represented by the angel of death, the power of evil, or Satan, has extremely pejorative connotations which it does not have in Indian or Chinese religious philosophies where cosmic libido, having no such direct political associations, is seen as complementary rather than antithetical to cosmic consciousness.

Using modes of livelihood as a criteria which have both political and evolutionary associations, the religious philosophy of the Hebrews excludes both farmers and hunters from sanctioned inclusion in the collective of the chosen people. By virtue of the varying summations of collective experience which these modes of livelihood represent, the convoluted ways in which ego-negating and libido-abdicating philosophies of awareness are achieved exclude those groups and individuals who do not share in precisely the right kind and combination of collective experience, both conscious and unconscious. The nature of groups that can be included in the collectivity of the 'chosen people' are those who possess herding, rather than either hunting or farming, as a mode of livelihood. In addition, the myth involving the conflict between the wives of Abraham and their two sons, Isaac and Ishmael, makes a further differentiation. Ishmael is doomed, by virtue of the fact that his mother is a concubine of Abraham, rather than a wife, to acquire a nomadic mode of livelihood in the desert. Abraham's wife Sarah is seen to have a special direct relationship with the personalized God by virtue of the fact that in her old age He bestows a son, Isaac, upon her, in answer to her prayers.

The prominance of the 'female type' in the personage of Sarah, represents the 3rd efflorescence mode where Abraham's concubine, no more than a servant in Abraham's household, represents the 2nd efflorescence mode. The suggestion is that the harshness and severity of nomadic modes of livelihood suggest an evolutionary mode closer to 2nd efflorescence society than 3rd, and therefore, by the moral standards promulgated in the Old Testament, inferior, and thus not worthy of inclusion in the collectivity of the 'chosen people.' Thus, in Exodus, where the Hebrews are forced to acquire a nomadic mode of life when they wander in the desert for 40 years after their exodus from Egypt, this mode of livelihood is seen as being imposed on them as a form of punishment for their reversion to the hedonistic habits they enjoyed when they were enslaved in Egypt. They reverted to these habits when Moses went to commune alone with God on Mount Sinai and they were left temporarily leaderless. Their lack of faith in both Moses and the religious philosophies he promulgated to them resulted in their being punished in this way, signifying their temporary exclusion from the collectivity of the 'chosen people.' Nomadic mode of livelihood viewed as a punishment expresses the fact that such a relatively primitive mode removes the collectivity from the summed experiences that allow it to achieve a direct relationship with God, in the guise of the ego-negating and libido-abdicating religious philosophies of the Old Testament.

In the course of the expository narrative of the Old Testament and as the Hebrew collective attempts to establish itself as a great nation, with all of the collective modes of 4th efflorescence society that that implied, the contradictions between the basic religious philosophy of the Old Testament and national aspirations unwind a series of national misadventures and tragedies. The series of exclusions from the collectivity of the chosen people include not only hunters, farmers and nomadic wanderers, but also large urban city dwellers, a theme which appears as early on as Genesis in the myths and allegories of the exodus of Abraham and Lot from the large urban centers of Sodom and Gommorah. These large urban centers are seen as festering enclaves of immorality and sin which prohibit the 'chosen people' from following this mode of settlement and the livelihoods emerging from them. The association of large urban centers with libido aggression no doubt reflects the earlier experience of the enslaved masses of Egypt in the urban centers of the Egyptian empire where they were so brutally oppressed. Other urban centers outside of the Egyptian empire in Near Eastern civilization symbolized by Sodom and Gommorah, and Ur, the urban civilization from which Abraham earlier exiled himself because of the moral corruption that apparently was seen to exist there as well, were politically independent of the Egyptian empire and probably represent the established empires of the Babylonians and Sumerians from about 5-8000 years ago.

Idol worship which is continually condemned in the Old Testament refers to the habits of these Near Eastern empires derivative of the Egyptian which personalized cosmic libido not in the person of the Pharaoh or King but in animal forms, particularly the bull. Acknowledgement of cosmic libido in any form was anathema to the Hebrews because of its close association with the tyranny of the Egyptian Pharaoh who believed himself to be its personification. When Abraham and Lot leave Sodom and Gomorrah, which are to be destroyed by the wrath of God, all of Lot's relatives are warned not to look back on the cities as they are destroyed, else they will be turned into pillars of salt, which is what happens to Lot's wife when she disobeys this injunction. The transformation of human beings into pillars of salt suggest the mummification of dead Pharaohs, where flesh was treated with chemicals to ensure the maintenance of its shape and form in time long after death.

The first and greatest King of the nation of Israel, David, is most pointedly a poor shepherd, who does not come from an urban environment unlike the manic-depressive King Saul, the very model of 2nd efflorescence man, whom he replaces. When the Kingdom under David and his son, Solomon, reaches the proportions of a great nation ruled by large urban centers, evil and mischance continually befall the nation, evil and mischance continually predicted by a series of prophets all from rural regions, where herding is the mode of livelihood. It is in this mode that even Jesus Christ is cast when he comes from the rural Nazareth to throw the money changers out of the temples in the urban center of Jerusalem. However, Jesus is above all unacceptable to the priests of Israel because he denies the series of exclusions which prevent all but those who share the particular collective past of the Hebrew nation, in all of their vicissitudes, historical, psychological, social, political, and economical, with both the direct and indirect influence of the Egyptian empire, from joining the collectivity of the 'chosen people.'

From a historical point of view Christ's message is more important in what he wishes mankind, especially the Hebrews, to forget rather than in what he wishes them to remember. His recurring theme of forgiveness and mercy on a political and historical level is a call to obviate the past, in the interests of the future, and in this call for obviation of the past he also obviates the series of carefully defined exclusions which prohibit most of mankind from joining the collectivity of the 'chosen people.' The figure of Christ not only gives a clear form to God, as exemplified in his own person as the son of God, but also a clear form to Satan, the personification of evil whom he personally confronts and whose existence he acknowledges. The personification of evil in Satan is carefully evaded throughout the Old Testament because of the necessity such personification implies in recognizing cosmic libido. Due to the direct political and historical experiences of the Hebrews with the oppressiveness of Pharaonic rule with whom cosmic libido is so closely associated, the personification of evil in the guise of Satan is largely ignored in the Old Testament, althought it is occasionally eclectically alluded to.

By personifying evil directly in the form of Satan, Christ allows mankind to recognize and confront the substance of evil and thus affords mankind the opportunity to overcome it, no matter what particular shape or form it manifests itself on any phenomenological level. By focusing unpleasant collective experience in a substantial rather than a particular way, Christ encourages the development of not only cosmic consciousness but also rational consciousness. With cosmic consciousness and cosmic libido in proper focus, rational consciousness can be developed and provides a bridge between modes of late 3rd efflorescence society and 4th efflorescence proper.

Chapter XII

GREEK AND INDIAN CIVILIZATIONS: EVOLUTIONARY ADAPTATIONS IN TWO TYPES OF OBJECTIFICATION MODES

The religious philosophies of the so-called 'underdeveloped' world, especially of India and China, have few mythic components as such, having long been refined into more or less systematic philosophical systems of a religious nature that are the fullest expression of modes of the late stages of the 3rd efflorescence. For those civilizations such as the Hebrews and the Greeks whose religious philosophies are contained in mythic form, it is because the origins of the civilization, as historical experience, were still fresh in collective memory, and as such could still be expressed in mythic form which is a basically dramatic mode, full of romance, adventure and tragedy. While the Old Testament narrative is mythic and hence historical, it is also an expression on many levels at once of other phenomenological experience apart from history, such as the processes of evolutionary development and religious philosophy of an ego-negating and libido-abdicating nature. This multi-level expression of the Old Testament is also a result of the not too distantly removed subjectification of phenomena as a mode of thought and conceptualization characteristic of the people who authored it. For the Greeks, however, who were entirely in a phase of objectification of phenomena, myth was a vehicle by which they narrated the diverse experiences of their wandering race, and this experience included the dutiful documentation, in a rather matter of fact way, of the various myths and beliefs they encountered among the Near Eastern and Asiatic peoples with whom they came into contact in the course of their wanderings.

Although the Greeks were in a phase of objectification of phenomena while the Hebrews were only partly so, both the Greek and the Hebrews reflected a rather direct experience with the Egyptian empire and its aftermath, although for the Greeks this experience was more in the fashion of being able to document this experience at second hand, rather than having shared in it in terms of either actual collective conscious or unconscious experience. In the Greek myths all divine personages are personified, including the original creators of all biological life, Heaven and Earth. Heaven was seen as the father of life and represented cosmic consciousness while Earth was its mother and represented cosmic libido. The transformation of cosmic libido, cruel, brutal and fertile to a female personage reflected the ease with which Greek society was already at that time entrenched in 4th efflorescence modes, and where there was no need to deify or idealize superego, which was so well-entrenched and thoroughly institutionalized that it needed little ideological reinforcement. The Titans, who were the children of Heaven and Earth, all clearly represented, in their personification of strangeness and gigantic size, the apt brutality, monstrosity and aggressiveness of 2nd efflorescence modes. Their father, Heaven, who represented cosmic consciousness, was so appalled at their uncivilized character that he attempted to imprison these all-masculine children, to prevent their libido aggressiveness from wreaking havoc on nature and in general from offending his moral sensibility. He left the least offensive of his many children at large, including the one-eyed Cyclops, who symbolized the myopia of 2nd efflorescence modes, and the more powerful and well-formed Titans, among whom was the Titan, Cronus, and the Titan, Prometheus.

Mother earth, representing cosmic libido, conspired with Cronus to revenge against her husband, and Cronus was thereby able to produce his own progeny, who included the Giants, another race of monsters; the Furies who were satanic in aspect and whose function was to punish sinners; and the Gods themselves. Cronus, who ruled with his sister-Queen, Rhea, with whom these offspring were produced, had definite Pharaonic associations. By a form of magic and divination, Cronus learned that he too, in true 2nd efflorescence form, would be deposed by one of his sons, so he ate them as they were born, depicting in graphic form the brutal parenting characteristic of 2nd efflorescence modes and suggesting as well the kind of paternal treatment the enslaved masses of Egypt could expect from their Pharaohs.

Disguised as a stone by his mother, Rhea, Zeus escaped his father's ravenous appetite for libido aggression and lived to over-throw him with the help of his uncle, one of the remaining and unimprisoned Titans, Prometheus. The structure of this myth is almost entirely similar to the Hebraic myth of Moses, who is saved from Pharaoh's wrath by Pharaoh's daughter, who was presumably, as well, his Queen, and raised in Pharaoh's house, illicitly, as his own son or grandson. Like the later Greek myth of Zeus, Moses lived, if not to overthrow the Pharaoh entirely, to establish himself as a leader of a separate people in rebellion against and independent of Pharaoh's rule. Both the myth of Moses and the myth of Zeus echo an even earlier Egyptian myth of the conflict between Osiris and his son Horace and Seth, Osiris' brother-in-law and cousin, even perhaps brother. The recurrence of these myths of titanic struggles between personages of God-like powers who are closely related by blood no doubt reflects the historical recurrence of such struggles among the royal households in the Egyptian empire. The denouement of these myths which result in the establishment of new dynasties both inside and outside of the Empire proper reflects a fusion of many such historical incidents in a single set of symbols and structural elements. Prometheus, the 'good' Titan, was uncle to successful Zeus, in the same way that Osiris was emasculated father to the successful Horace. While there is no such explicit royal avuncular persona to aid Moses in the corresponding Hebraic myth, his brother Aaron, who was better able to communicate with the Pharaonic elite than Moses himself, may represent just such a disguised royal avuncular figure.

In addition, in the Old Testament there are 'good' Pharaohs in avuncular roles such as the Pharaoh who assisted Joseph against the libido aggression of his brothers, and of course the daughter of Pharaoh who was Moses' adoptive mother stands in the same helpful kinship relationship to Moses as Isis to Horace, and Rhea, Zeus's mother, to Zeus. Just as the good Pharaoh puts the skills of Egyptian civilization at the disposal of Joseph, and Moses' Pharaonic adoptive kin presumably do the same for him, so does Prometheus not only aid Zeus in his struggle with his father by enlisting Titanic powers in his aid, but also later brings the gift of fire to mankind, wreaking the displeasure of Zeus upon him in this aspect of the Greek myth, where Prometheus is chained to a rock, his liver devoured by vultures for his pains on behalf of the human race. This aspect of the Greek myth also echoes not only the Hebraic myths but also the earlier Egyptian myths, where Osiris suffers dismemberment so that his son, Horace, whose eye is also lost in the titanic struggles that ensue, may ultimately prevail over Seth.

In the Hebraic myth, the snake may be a disguised surrogate for the dismembered Osiris in the Egyptian myth and the ravaged Prometheus in the Greek myth, where all three figures, the Hebraic snake, the Egyptian Osiris, and the Greek Prometheus suffer in the cause of enlightenment for mankind. The difference between the Egyptian and Hebraic myths on the one hand and the Greek myths on the other hand is the relative clarity of the Greek myths which the others lack, due of course to the modes of objectification of phenomena in Greek thought and conceptualization. In the Greek myths, the personae are all clearly symbols, idealized and mythicized Gods who have no direct relationship to human existence and do not at all share in the experience of the human collective. In the Egyptian myths the personae are living Gods, that is, both humans and Gods at the same time, and each Pharaoh who ruled identified himself directly as the reincarnation of the living God, Horace. In the Hebraic myths the personae are actual individuals in an historical drama, and at the same time are meant to represent the manifestation of the one living God, as testified by their privileged channels of direct communication with Him.

The entire pantheon of Greek Gods are entirely anthromorphicized in a way that is completely alien to Egyptian, Hebraic and other Near Eastern myths, again due to the mode of objectification of phenomena that characterized both Greek thought and language. The Greek Gods created the races of men, in five stages of metallic substances, beginning with gold and ending with iron that is the present race of mankind. This depiction of evolutionary history in such a matter of fact manner, where the race begins in primitive naturalness in the 1st efflorescence and undergoes many vicissitudes including a 'golden age' which reflects peak 3rd efflorescence society where libido aggression and ego-recognition lie dormant, is another example of both the mode of objectification of phenomena in Greek thought and also the systematic and collative characteristics of Greek legends, where knowledge was picked up from other races with whom the Greeks came into contact in their wanderings and duly incorporated into a clarified logos of myth. This logos was in effect an encyclopaedia of myth collected from other races.

Invariably the pantheon on the Greek Gods all had humanized characteristics. Each God or Goddess of the family of Gods, headed by the father Zeus and the mother Hera, had particular attributes and represented certain human ideals. While none of these Gods or Goddesses were regarded in any way as perfect, altogether as a family and as individuals they were a far more congenial lot than the primitive family of Titans who preceded and engendered them. They could not and were not in any way identified with primal psychic forces, as these forces were clearly focused in the creators of biological life, father-Heaven as cosmic consciousness and mother-Earth as cosmic libido. Ego recognition and superego were focused in the immediate progenators of the Gods, the Titans who were the children of Heaven (cosmic consciousness) and Earth (cosmic libido). Here ego-recognition had cosmic associations in the overwhelming ego of the Titan, Cronus, as superego had cosmic associations in the eternal and abiding sacrifice of the Titan Prometheus who allowed himself to be pilloried by Cronus or ego for the welfare of mankind.

The Gods were idealized humans in 4th efflorescence modes, and as such were characterized most consistently by their possession of rational consciousness. This rational consciousness endowed in them a powerful sense of justice which, idealized by humans, could be employed in the service of rendering decisions, just, rational and wise, in every aspect of individual and collective existence. The distance between men and Gods was seen as great, but not too great, since like men, Gods suffered under the constraints of the universalized psychic forces of ego, superego, consciousness and libido, all of which were summed up in the Greek concept of 'fate.'

From a historical point of view, the pantheon of Gods represented a *summa* of Greek collective experience, that there did indeed exist in certain times and places political and social elites whose power, knowledge, and accessibility to every conceivable kind of luxury was so disparate from the peoples over whom they ruled that they could indeed be characterized as 'Gods.' That such a relatively pleasant face was put on these elites stemmed from a number of factors. For one thing, the Greeks themselves as a collective had never actually been subject to this kind of elite, so they were able, from a distance, to view this kind of elite in a positive manner, something which the Hebrews and other Near Eastern peoples were totally unable to accomplish. For another thing, given the rich and wide diversity of their wanderings in the late stages of neolithic civilization, it is possible that the Greeks encountered relatively benevolent elites of this kind in kingdoms established far from the ken of primal Pharaonic rule, perhaps in the more successful phases of the Hebraic kingdom, or in India, or elsewhere. Also, due to the rational philosophy that they themselves evolved over time, it was inevitable that they would project such a pleasant face on the Pantheon of their own Gods, who, after all, were always to them never actually people, but human ideals or humanized spiritual entities.

As a collective who had come full fledged into the 4th efflorescence, the myths of the Greeks reflect vestiges of the 2nd, 3rd, and 4th efflorescence, where 2nd efflorescence modes were mitigated by the entrenched superego of the 3rd, and where both 2nd and 3rd efflorescence modes were organized in terms of the collective ego-recognition and libido aggression of the 4th. That both 2nd and 3rd efflorescence modes were extant and almost taken for granted, where the collective had already attained the higher stage of the 4th, can be seen from the much quoted lines of Aristophanes on the Greek myth of creation:

. . . . Black winged Night
Into the bosom of Erebus dark and deep
Laid a wind born egg, and on the season rolled
Forth sprang love, the longed-for, shining with
wings of gold.

The 'egg' is an expression of reptilian or masculinized reproduction, signifying 2nd efflorescence modes, while the creature or entity of 'love' which it gives birth to signifies 3rd efflorescence modes of institutionalized superego. The particularization of characteristics ascribed to the various Greek gods represent as well the melange of 2nd, 3rd, and 4th efflorescence modes that exist in a society that is properly in the 4th efflorescence. While one God, Apollo, is significant for his characteristics of reason, another, Mars, is significant for his warlike characteristics, while the Goddess

Athena represents justice and Venus represents romantic love, and so forth. No one God or Goddess embodies all of the characteristics valued by the society, as does Ra in Egyptian myths, or Maladuk in Babylonia, or Jehovah in Hebraic myths. No one God or Goddess is entirely good or entirely evil, and none are either omnipotent or omniscient. This congenial lot of idealized humans, full of foibles and prone to misadventure in the face of fate, reflect the social and political perception the Greeks had of themselves, as individuals and as a collective.

It was no doubt this perception the Greeks had of themselves, as individuals and as a collective with mixed characteristics whose differences could only be adjudicated by the exercise of rational consciousness, that led to the development of democracy as a social and political mode among the early Greeks. Their notions of cosmic libido and cosmic consciousness, although present in their myths and their culture, and summed up by the concept of 'fate' or destiny, were not deemed important enough to be subject to serious rational analysis or clung to with any sense of religious fervour. In Thucydides' famous literary and philosophical history, *The Peloponnesian War*, which documents in a cause and effect fashion the vicissitudes of the wars between Athens and Sparta about 500 B.C., Thucydides subjects the causes of these wars and the ultimate decline of the Greek empire to rational analysis. The headlong and casual rush into 4th efflorescence modes by a collective is nowhere more poignantly expressed than in the characterization of the Athenians by the Corinthians at the council of Sparta. As a collective squarely in the 4th efflorescence mode, Thucydides contrasts the Athenians with the Spartans, who are more hesitant about adopting all the modes of 4th efflorescence society and remain to some extent attached to modes of the late stages of the 3rd efflorescence:

> The Athenians are addicted to innovation, and their designs are characterized by swiftness alike in conception and execution; you have a genius for keeping what you have got, accompanied by a total want of invention, and when forced to act you never go far enough. Again, they are adventurous beyond their power, and daring beyond their judgement, and in danger they are sanguine; your wont is to attempt less than is justified by your power, to

> mistrust even what is sanctioned by your judgement, and to fancy that from danger there is no release. Further, there is promptitude on their side against procrastination on yours; they are never at home, you are never from it: for they hope by their absence to extend their acquisitions, you fear by your advance to endanger what you have left behind. They are swift to follow up a success, and slow to recoil from a reverse. Their bodies they spend ungrudgingly in their country's cause; their intellect they jealously husband to be employed in her service. A scheme unexecuted is with them a positive loss, a successful enterprise a comparative failure. The deficiency created by the miscarriage of an undertaking is soon filled up by fresh hopes; for they alone are enabled to call a thing hoped for a thing got, by the speed with which they act upon their resolutions. Thus they toil on in trouble and danger all the days of their life, with little opportunity for enjoying, being ever engaged in getting: their only idea of a holiday is to do what the occasion demands, and to them laborious occupation is less of a misfortune than the peace of a quiet life. To describe their character in a word, one might truly say that they were born into the world to take no rest themselves and to give none to others.

In the end, however, even fate is not called upon to explain the tragedies, and misadventures which lead to the decline of the Greek empire by Thucydides. Instead, Thucydides ascribes these causes to an incomplete utilization of rational consciousness by Greeks on both sides which prevents collectivity of action, the fullest expression of 4th efflorescence modes. In an uncanny fashion Thucydides is able to predict the rise of Roman civilization when he ascribes the critical defeat of the Athenians in Sicily to the greater ability of the Sicilians to exercise collective action than the Greeks themselves. Since Thucydides himself was an exiled Athenian Admiral living in Sparta, this overwhelming faith in the ability of rational consciousness to solve all individual and collective problems is treated with a certain sceptical sensibility, but even this scepticism, which is in the mode of the later Roman, pre-Christian stoicism, is unexplained and is similar to the sense that those causes for man's misadventures and tragedies in the 4th efflorescence outside of rational miscalculation can be ascribed to 'fate,' as in the earlier Greek tragedies. In fact, unlike the Hebrews and other Near Eastern and Far Eastern peoples, the Greeks' encounter with 'fate' was never so traumatic as to force them to come to terms with it in any deep philosophical or religious way. Because of the relatively untroubled historical experience of the Greeks, collective consciousness was not split from collective unconsciousness. In the memory of the Greek race, tragedy and misadventure could be clearly visualized and could largely be seen as the result of rational miscalculations that could, at least in retrospect, be rectified. Historical experience had led them to believe that they and they alone were responsible, on the human and historical level at least, for their own fate, and their Gods and fate itself were very largely a reflection of their own happy historical experience.

By contrast, in the religious philosophies of the Far East, like those of the Near East, there is a dread association with collective action that is not explicable by rational causes, and there is normally no attempt to apply purely rational cause and effect analysis to historical experience. In the Indian *Bagavad Gita*, the dread association with war and political struggles, as aspects of collective action, is so overwhelming that the hero resolves his anguish by denying the reality of collective action, and thus its significance. The sense of dread association with free-ranging collective action in both Indian and Chinese philosophy, however, cannot be traced to any specific sets of historical experience, as in the case of the Hebrews and other Near Eastern peoples, for unlike both the Ear Eastern peoples, the Hebrews and the Greeks, neither the Indian nor the Chinese express their religious philosophies in a logos of historical myths. In fact, the Indian, and to some extent early, but not late, Chinese written culture can be seen as anti-historical. Both the *Upanishads* and the *Bagavad Gita* deny the validity of myth as a substantial expression of their religious philosophy or even their collective experience. Like all commonplace reality, myth, as an expression of history, is assigned to the realm of the profane, the unreal and the irrelevant. Where myths exist in Indian culture they are valued only in the sense that they can be utilized as a form of 'sacrifice.' 'Sacrifice' as the way of 'yoga' implies that myths are important only to be denied, like all other modes of commonplace reality. To 'sacrifice' or abandon commonplace modes of reality is an expression of the ego-negating, libido-abdicating religious philosophy of the Indians, just as the

Chinese *Tao* advises a constant process of 'unlearning' commonplace knowledge of all kinds.

In the Indian philosophy, 'sacrifice' of all commonplace reality, and action past and present, suggests that the only way reality or action is valid is if it is identified in a total way with cosmic libido. That is, actions of the individual or the collective are never self-impelled, but are only a facet of the actions of cosmic libido. Through identification of all action, individual and collective, with cosmic libido, cosmic consciousness, or Brahman, Atman, OM, can be attained. In this mode of consciousness, both individual and collective, ego-recognition is entirely lost, and a pure cosmic consciousness, beyond the rational consciousness that derives directly from ego-recognition, is attained. The description of such consciousness is given in the *Chandogya Upanishad*:

> This is the spirit that is in my heart, smaller than a grain of rice, or a grain of barley, or a grain of mustard-seed, or a grain of canary-seed, or the kernel of a grain of canary-seed. This is the spirit that is in my heart, greater than the earth, greater than the sky, greater than heaven itself, greater than all these worlds. This is the spirit that is in my heart, this is Brahman.

Although there is a fundamental identity between the *Brahman* of Indian philosophy and the *Tao* of Chinese philosophy, there are critical differences in the nuances of these two great Far Eastern civilizations due to the differing historical circumstances which produced them in their evolved form. The Indian collective was a melange of basically two entirely different groups. From the east and south, the Vedic peoples were the descendents of the Near Eastern peoples who had shared directly in experiences of the Egyptian empire and who brought with them to the Indian subcontinent all the skills of civilization they had acquired in their wanderings from the neolithic civilization that extended throughout the Mediterranean and southwest Asia during the long period of Egyptian decline.

Since the distance from India to Egypt is much greater than from the Near Eastern regions to Egypt, it is understandable that Indian civilization developed later in historical time than Babylonian, Sumerian, and Hebraic civilizations of the Near East, as archeological evidence will attest. However, at the same time as the Indian subcontinent was receiving the influx of Vedic type peoples from the south and the east, from the north and the west the Indian subcontinent was being invaded by Nordic hordes of Aryans who found both climactic and ecological conditions more favourable than the inhospitable regions which had engendered them, and also a vast potential for pillage among the Vedic peoples who had already settled in the subcontinent and who possessed all of the skills and riches of late neolithic civilization. Nonetheless, the invasion of the Indian subcontinent and the pillage of the Vedic peoples took place under entirely different circumstances from the earlier invasions of Nordic hordes among the peoples of the Nile valley some 15,000 years earlier.

The Vedic peoples of the Indian subcontinent were far removed from the inhabitants of the Nile valley in those earlier times who had just peaked 3rd efflorescence civilization. The Vedic peoples possessed the entire historical logos of both direct and indirect experience with the Egyptian empire, and they possessed this experience as a matter of both collective consciousness and unconsciousness. Having established their independence, both socially, politically, and geographically from conditions of enslavement under Pharaonic rule and its later vestiges in Near Eastern civilization that was an unhappy combination of 2nd and 4th efflorescence modes, they in no way resembled the innocent and unsuspecting peoples of the Nile valley who had allowed their newly established institutionalized superego to become the vehicle of brutal political and social oppression. However, unable as they were, by virtue of remaining largely in late 3rd efflorescence modes, to combat the Nordic Aryan invaders by force, for thousands of years they employed their energies in absorbing these invaders through the sophisticated psychological and religious mechanisms at their disposal.

Among the Nordic Aryan invaders of the subcontinent, powerful vestiges of 2nd efflorescence society were combined with emerging modes of the 4th, while 3rd efflorescence modes were vague and ill-formed. Superego was not broadly institutionalized but was strongly attached to beliefs in superstition and magic without any well-organized religious themes. While for obvious reasons the Vedic peoples were not about to make the same mistake as the original inhabitants of the Nile valley and share with their Nordic conquerors the summation or foundation of their skills of civilization, they blended certain aspects of their own history in mythic form with the more meagre mythological resources of their invaders. In doing so they were able to convince the invaders that both their own myths and those of the invaders were irrelevant to an acquisition of the skills of civilization, which of course they were not. While identification of libido in any form, individual, collective or cosmic, had become taboo among Near Eastern peoples such as the Hebrews because of the associations that libido in any form had with the painful experiences of enslavement under Pharaonic rule, this was a luxury that could be ill-afforded by the Vedic peoples on the Indian subcontinent, who were constantly confronted with the overwhelming ego-recognition and libido aggression of Nordic Aryan invaders.

Libido was brought to consciousness by the Vedic peoples in order to focus the expression of libido aggression of the Aryans, at which point it was immediately transformed from individual and collective libido aggression to cosmic libido aggression. This was accomplished through the devices of religious and magical associations which the Aryans dreaded on the basis of ill-formed superego that treated all such associations in the form of superstition. Thus the Vedics permitted the Aryans to channel their powerful libido aggression into religious modes. Libido aggression which demands a goal was given the most difficult goal of all: abdication of libido aggression. Ego-recognition as a powerful psychic force among the Aryan invaders was similarly channelled, and in such a way that it changed the nuances of the religious philosophy of the Indian subcontinent and made it substantially different from related religious philosophies in the Near East.

Whereas in religious philosophies of the Near East, such as the Hebraic, the relationship between the individual and God, or cosmic consciousness, was direct, in Indian philosophy it became indirect. The individual was only able to have a relationship with cosmic consciousness through the intermediary of a religious leader, guru, and later Buddha. This leader, guru, or Buddha was seen as an exemplary, and did not embody cosmic consciousness as the Pharaoh did, or as a derivative type of King or surrogate Pharaoh might. The leader, guru, or Buddha had found a way, through the sacrifice of myth and action, and all commonplace reality, both Vedic and Aryan, of identifying totally with cosmic libido. In this way he had, as an individual, found the way or 'yoga' by either objectification or subjectification of phenomena modes. His way of acting served as an example so that other individuals, as his followers or students, might emulate his behaviour and in that way also attain cosmic consciousness. The overwhelming ego-recognition as a psychic drive had found a channel, as ego must first be recognized before it could be negated, and as an ego-goal no greater difficulty existed than the negation of ego itself.

The guru or Buddha as the example of ultimate ego-recognition through ego-negation did not demand that followers subsume their own ego-recognition in that of the leader, as in Egyptian civilization. On the contrary, the goal of every follower was to become a guru himself, or a leader, and when he had achieved the same state of ego-negation as the guru, he attained the same inimitable social status as the original leader. In this way individuality of the 2nd and 4th efflorescence modes was channelled into the non-individuality modes of the late stages of the 3rd efflorescence. Powerful psychic forces of ego-recognition and libido aggression characteristic of 2nd and 4th efflorescence modes were channelled and transformed into the ego-negating and libido abdicating philosophies of the late stages of the 3rd efflorescence. Myth and history, Vedic and Aryan, were left behind and forgotten, Vedic because it was regarded as politically dangerous in the hands of the Aryan invaders, and Aryan because it was regarded as superficial and irrelevant in the context of the more advanced Vedic civilization.

The development of a highly structured caste system in Indian civilization, as well, reflects the fusion of conflicting modes from Vedic and Aryan collectives which produced Indian civilization as it still exists today. The hierarchical nature of the caste system provided another channel for the ego-recognition and libido aggression of the Aryan tribes who were being absorbed into early Vedic settlements, and allowed a social and political method of adjudicating the differences between the natural proclivities and aims of these two different collectives. The Brahman caste at the peak of the hierarchy represented the original and creative fusion of Aryan and Vedic modes in the institutionalization of a leader/guru caste or class. Diplomatically, the second rank of warrior caste was made available to the more aggressive Aryans, who, however, were never allowed to achieve supremacy in Indian civilization due to the innovative, reigning guru/leader caste. The lesser ranks of merchants and farmers provided a protective niche for the masses of Vedics who retained the skills of civilization they had brought with them from their earlier existence in Near Eastern regions, and the caste system, although it provided them with a rank lesser than the Aryan warriors, allowed them to isolate their knowledge from the aggressive Aryan invaders in whose hands it might otherwise have been used for offensive political, social, and religious purposes. The institutionalization of a pariah or untouchable caste was a dramatic way of illustrating the libido aggressiveness of the entire caste system, providing an indisputable and easily expressible mechanism for the powerful drives of ego-recognition and libido aggression of the Aryans on a collective, as well as an individual, level.

Thus, while the particulars of history expressed in myth were discarded in Indian civilization, history was expressed in the very structure of the society. As a matter of expediency, the processual aspects of history were discarded in favor of the substantive. From the Vedic point of view, this was an advantageous exchange, because of the fear they had of transmitting their own historical logos to the more aggressive Aryans. From the Aryan point of view, it was also advantageous, as they had not developed a mode of conceptualization in tandem with the sophisticated subjectification of phenomena necessary to understand and utilize the skills of civilization made available to them by Vedic civilization. Thus, for entirely different reasons from the Greeks, Indians civilization nonetheless adopted objectification of phenomena as a mode of thought and conceptualization. In the late establishment of Vedic civilization on the Indian subcontinent, the Vedic peoples had been well exposed to the tail end of neolithic civilization in Near Eastern regions and had at their disposal phonetic written language and particularized language chains as a mode of thought and communication, under which of course existed the substratum of subjectification of phenomena modes of thought and communication. In contact with the Aryan invaders, this subjectification of phenomena substratum was discarded. For those Aryan invaders who had never possessed such a substratum as a fact of collective historical experience, acquisition of objectification of phenomena modes in spoken and written language was easily and gratefully learned, representing as it did a distinct improvement on their own rudimentary written and spoken language forms, which were an expression of ill-formed subjectification of phenomena and totally inadequate to the mastery of the skills of civilization they encountered among the Vedic peoples.

The rationalization of the Indian collective, from the point of view of societies that have come fully into the 4th efflorescence, seems untenable due to the caste system, even though, in reality, 4th efflorescence societies have hierarchical structures less extreme only in degree, disguised in terms of economic or racial class systems, or in euphemized systems of institutionalized slavery in the guise of economic or political colonies. In particular the existence of an untouchable caste is offensive to 4th efflorescence sensibilities since it obviates the opportunity for that caste to rise into the higher echelons of society, on a tutorial basis, so dissimilar from the open-ended model of class and social mobility characteristic of all 4th efflorescence modes. The lack of an open-ended social and political mobility model in Indian society is in some ways a caveat, betraying as it does the inability of individuals in 4th efflorescence modes to comprehend and acknowledge societies in the late stages of 3rd efflorescence modes as viable social entities with their own kind of special integrity.

The caste system is an open and honest acknowledgement of the ego-recognition and libido aggression of both 2nd and 4th efflorescence modes which it is the object of modes of late stages of the 3rd efflorescence to obviate through their religious philosophies. It is allowed to exist because it permits expression of ego-recognition and libido aggression as real psychic forces which are, however, viewed as essentially immoral. The existence of the lower castes, particularly the untouchables, in their misery and degradation represents aspects of cosmic libido as expressed in collective and individual forms. Thus, in its direct association with cosmic libido, this class of pariahs is indeed 'untouchable' as it represents, as a collective expression, the action of the cosmic mind or God. For good or for evil in commonplace reality, as human collective and individual existence represents, the actions of the cosmic mind, like the cosmic mind itself, are 'untouchable'–that is, they are not a proper subject for human judgement and in fact are deemed inaccessible to rational analysis. From the point of view of the religious philosophies of India, Indian society is in fact as open-ended and mobile as 4th efflorescence society is from the point of view of its rationalized philosophies.

Within the ken of Indian religious philosophy, it is possible for a member of any caste, including the untouchable, to become, through mastery of this common religious philosophy, a guru/leader or Buddha, and thus to attain to the highest social status in Indian society. Many modern political and social leaders of Indian society, like many Buddhas in the historical past of Indian civilization, have come from the 'untouchable' class, and there is no more sense of contradiction in Indian society about the rise of an individual from the poorest and most despised social caste to a position of political and social prominence than there is, for example, in modern American society in the 'rags to riches' or Horatio Alger model. While in modern American society such mobility is judged by the criterion of the mastery of civilization skills, particularly economic ones, in Indian society such mobility is judged by the criteria of mastery of ego-negating and libido abdicating skills which allows the individual to become exemplary in those skills, and thereby, from an outside point of view, to gain enormous ego-recognition and even potential for libido aggression.

In addition, the theories of reincarnation which co-existed with the religious philosophies of India and were no doubt a vestige of early Vedic collectives, allowed for 'democratic' models of social and political mobility. According to these theories, every living creature contained a quintessence of the absolute self, Brahman, Atman and OM, and depending on the particular theory of reincarnation, many different ones being prevalent, either by accident or due to the mastery of ego-negating, libido-abdicating religious philosophies, each living creature could be reincarnated, in its next life form, into a higher level of existence. In some theories a thing as lowly as an insect could be reborn as a man into a member of the warrior or even the Brahman caste. This accounts for the prohibition in certain Indian sects against not only the eating of meat, but even the killing of any living creature, including insects, for that which is inadvertently slain might contain the special quintessence of the Absolute Self or cosmic mind of

God which, in the next cycle of death and rebirth, will be manifested in commonplace reality in the form of a higher life form, or member of a superior Indian caste.

In some of the variants of Buddhism, which attempted to rationalize and systematize the religious philosophies of India as they had developed over thousands of years on the Indian subcontinent, reincarnation of human individuals into a better station of life than they had previously enjoyed was related to the degree of mastery that they had evinced in the former lifetime in ego-negating, libido-abdicating modes. However, all manifestations of commonplace reality, no matter how favorable in terms of the political, social or economic conditions under which the individual lived, were considered a mode of suffering, alienating the individual as they did from absolute identification with the mind and actions of the cosmic consciousness. Thus, in Indian theories of reincarnation which envisioned an endless cycle of death and rebirth, the ultimate goal for the individual was the termination of this cycle, so that at some point he would no longer be reborn, rebirth under no matter how favorable the conditions being regarded as a form of punishment and suffering.

A complete mastery of the ego-negating, libido-abdicating religious philosophies of India, as exemplified in the historical personage of Buddha, represented the achievement of the termination of the abhorrent cycle. From a social and political point of view, these theories represented a 'democratic' and universally egalitarian view of individuals, regardless of caste. Any given, existing social station which an individual possessed was viewed, from the point of view of these philosophies, not in terms of the fulfillment of individual ego-recognition and libido aggression, but as the degree to which ego-negating and libido-abdicating philosophies had been mastered in a previous lifetime. Similarly, no matter how miserable or unfortunate the social status of any individual might be, it was not seen in terms of his lack of fulfillment of ego-recognition or libido aggression, and depending on how well he mastered ego-negating and libido abdicating philosophies under his present circumstances, he could look forward to enjoying a more favorable social status in his next reincarnation.

The survival of theories of reincarnation in India was undoubtedly testimony to the direct experience which the Vedic peoples had with the civilization of the ancient Egyptian empire. The survival of these theories of reincarnation on the Indian subcontinent also indicate that the original Vedic settlers of the subcontinent had a rather different kind of direct experience with Egyptian civilization than the Near Eastern peoples who derived from the Egyptian empire and disseminated neolithic civilization in that region. Among the Near Eastern peoples, typified by the Hebrews, there was an abhorrence of reincarnation theories and associations that bordered on the obsessive. To these peoples theories of reincarnation were associated directly with the political and social oppression that they had suffered at the hands of the Pharaohs, whose entire goal was to subjugate the people under their rule for purposes that they believed would enhance their own chances for reincarnation. Of all the myriad groups who emigrated or were exiled from the Egyptian empire, however, not all were necessarily members of classes that were totally enslaved.

In the Old Testament myth of Joseph, who, having become minister to Pharaoh, eventually brought all of his tribe to Egypt, the narrative claims that for many generations this tribe was highly favored by the Pharaohs, and enjoyed a status comparable to that of the Pharaonic elite itself. In addition, the establishment of the Egyptian empire in its length and breadth over tens of millenia necessitated the entrenchment of political and social elites widespread throughout the empire. Among those peoples who migrated to the Indian subcontinent from the Egyptian empire during the period 10-8,000 years ago and who were the prototypes of the later Vedic civilization, there were no doubt a large percentage of these favored political and social elites. This is suggested in the survival of reincarnation theories, where these same theories were treated in the most pejorative way among those peoples of the Near East who had established their independence from the Egyptian empire under the conditions of war and revolution and by means of force. Furthermore, once the migration of collectives originating in the Egyptian empire had established themselves on the Indian subcontinent, geographically remote from the power of the Egyptian empire, they were also no doubt able to look back, from their advantageous position, on many of the Egyptian religious and philosophical theories with the equanimity that such a safe distance from the Pharaonic elite might allow.

That the migration of these peoples to the Indian subcontinent, even in the first instances of the earliest migrations, was not the result of a happy relationship with the Egyptian empire can be seen from the fact that there was never any attempt to replicate the Pharaonic system or even the Imperial Egyptian system on the subcontinent. But the attitude of these peoples towards Egyptian civilization, unlike those in the Near East, seems to have been one of reform rather than revolution.

The theories of reincarnation were maintained, but not utilized in the interests of any particular political or social elite. They were reformed in order to permit the ego-recognition and libido aggression which reincarnation as a theory expressed in an over-weening and extreme fashion to extend to all members of the migrating collective, and not just to a few. The scientific and technological skills which had been employed in Egypt to ensure the physical reincarnation of the Pharaoh and his immediate favorites appear to have been entirely abandoned, and in this sense the negative reaction to the political and social modes of the Egyptian empire by the forerunners of Vedic civilization on the Indian subcontinent were pointedly expressed. The philosophical and religious concepts, which provided the basis for the theory of reincarnation and which had been largely suppressed or ignored in the Egyptian empire by the religiously and philosophically unsophisticated Nordic conquerors of that empire, were revived and extended. These theories provided the basis for the ego-negating and libido abdicating religious philosophies which evolved on the Indian subcontinent and which are differentiated from those in the Near East proper by the lucidity and clarity of their goals, free from the historical particulars and series of exclusions which preoccupy Near Eastern religious philosophies of the same ilk.

Reincarnation was seen as a necessary aspect of cosmic consciousness and cosmic libido, and, as one of the particular actions of cosmic mind and action in general, was entirely egalitarian and 'democratic' in its manifestation. As cosmic mind and cosmic action were entirely beyond the control of individual or collective will, and entirely outside the ken of rational analysis, no scientific or technological skill, however refined, could in any way affect its workings. For practical as well as philosophical and religious reasons, therefore, the scientific, technical and medical skills which had been developed under the Egyptian empire, at such great social, political and economic cost to the populace, to ensure the physical reincarnation of the Pharaoh and his favorites, were abandoned. The fundamental methods by which such skills were acquired remained in the collective consciousness and unconsciousness of the Vedic peoples on the subcontinent in sophisticated modes of both subjectification and objectification of phenomena until the encounters with the Aryan invaders of the subcontinent induced their almost total suppression. Reincarnation theory, that had originally been acquired under the circumstances of the Egyptian empire, in its reformed and egalitarianized form, was one more method by which the modes of 2nd and 4th efflorescence society of the Aryan invaders of the subcontinent could be absorbed and channelled into modes of the late stages of the 3rd efflorescence by the earlier Vedic inhabitants of India.

PART III

"AN ACCIDENT OF LANGUAGE: THE SPECIFIC DEVELOPMENT OF CHINESE CIVILIZATION IN THE FOUNDATIONS OF THE EGYPTIAN 'MASTER CODE' OR *I-CHING*"

Chapter XIII

LANGUAGE EVOLUTION EAST AND WEST OF THE EGYPTIAN EMPIRE

Of all the civilizations in the world whose dynamics are difficult to comprehend, at least from the point of view of 4th efflorescence society, none is more so than the Chinese. The modes of thought and behaviour of Chinese civilization often seem so remote from the point of view of 'developed' societies of the 4th efflorescence that they seem to belie any kind of common evolutionary and historical origin. From the point of view of the evolutionary and historical modes developed however, the relationship between modes of Chinese civilization and modes of both other civilizations in the late stages of 3rd efflorescence, and the 4th, can be understood. Like Indian civilization, Chinese civilization does not reflect either its early prehistorical past or its religious philosophy in terms of a coherent mythic structure like the Greeks, Hebrews and other Near Eastern peoples. But unlike Indian civilization which possess a tremendous logos of historical myths which it deliberately abandons as an aspect of its religious philosophies, Chinese civilization seems to contain few early historical mythic elements that are indigenous to that civilization.

The Anyang area around Peking in northern China, which is considered by Chinese historiography to be the original site of Chinese civilization, appears to have been settled as early or earlier than the Indian subcontinent some 10-15,000 years ago, but unlike the earliest Indian settlements, the early Chinese settlements were sparse and lacked the intellectual and cultural vigour of early Vedic settlements. If the evidence of a viable culture on the Chinese mainland is at about the same time as the Indian subcontinent, while the distance from China is both longer and contains far more formidable geographical barriers separating it from the center of Egyptian civilization in the Nile Valley, it stands to reason that in fact the settlement of China by collectives associated with the

Egyptian empire was even earlier than that of India. The sparseness of very early mythic literature in Chinese civilization is testimony to both the early date of settlement in China by collectives migrating north-eastwards, and to the sparseness of these settlements. While the very earliest settlements of the Indian subcontinent by privileged groups migrating, or in exile from, the Egyptian empire was a continuous process, where early settlement and privileged groups on the Indian subcontinent were bolstered in time by a stream of further settlements in contact with the neolithic civilization being developed in the Near East, for reasons of geography this was probably not the case in China.

Unlike India, reincarnation theories were never prevalent in Chinese civilization nor co-existed as an aspect of the ego-negating, libido abdicating religious philosophies of China. Unlike both India and the Near East, there is however tangible archaeological and historical evidence that tyrannical kings of China as late as 2500 years ago in the Chin dynasty believed that they could be spiritually, if not physically, reincarnated, and the system by which the king's household were slaughtered at his death and placed in the tombs with him, presumably to accompany him on his journey through the underworld to his rebirth, was an exact replica of the Pharaonic system of the ancient Egyptian empire. The earliest historical documentations of Chinese civilization in Chinese records are those of the Shang dynasty some 5-6000 years ago, and, according to these documentations and some additional archaeological evidence, the Shang dynasty represented a most tyrranical form of government which rivalled the Pharaonic rule in its brutality and oppressiveness.

The Shang dynasty is seen as of particular historical significance to the Chinese, since its overthrow in the form of violent revolution led to the establishment of the foundations of Chinese civilization as it survived into modern times, with the modes of social and political behaviour and the accompanying religious philosophies that characterize Chinese civilization in its fully evolved forms. It stands to reason that the Shang dynasty was not a single instance of a Pharaonic model of government and civilization in ancient China, but merely the last. Revival of this mode, in far more civilized and refined ways, occurred in the Chin dynasty 2200 years ago, and in many of the dynasties of post-historical

China the Emperor and the Imperial system, at least as symbols, were reminiscent of these earlier modes where the Emperor was regarded as the 'son of heaven' and thought of as a living divinity.

In terms of the early date of the settlement of China by collectives who had direct experience with the Egyptian empire, and by the geographical remoteness of China from Egypt, establishment of an Egyptian-derived civilization in China resembled that of India. However while the earliest settlers of the Indian subcontinent undoubtedly had direct experience with the Egyptian civilization, there was never in early India an attempt to emulate the system of government or the social and political modes of the Egyptian empire. On the contrary, both lack of evidence of such emulation, and the ease with which reincarnation theories were accepted and co-existed with the developing religious philosophies of India, suggest that the early settlers of the Indian subcontinent were elements of a privileged group in that empire who deliberately broke away from it. This appears not to have been the case at all in early Chinese settlement, where such settlement appears to have been a result of colonization of the Chinese mainland under direct Egyptian control and supervision.

The fact that Pharaonic modes of rule existed in China 5-6000 years ago, milennia after the demise and disintegration of Egyptian civilization, was no doubt a result of several critical factors. The most critical factor in the continued existence of Pharaonic modes in the colony long after the mother civilization had lost control over it was first of all due to the geographical remoteness of the colony in China from Egypt. Such geographical remoteness prevented the Chinese settlements from coming into contact with the series of revolts against the Egyptian empire in the Near East and which led ultimately to the dissemination of neolithic culture throughout that region. Separated as they were from these series of contiguous and continuous revolts that led to the spread of a more or less common neolithic culture in the Near East and parts of southwest Asia, the inevitable development of neolithic civilization with all of its particular modes was forced to occur independently in China, and as a result happened to spread later and much slower throughout the Chinese mainland than it did in areas contiguous to the Egyptian empire and where geographical barriers did not prevent its more rapid dissemination. Since neolithic civilization occurred so rapidly, so contiguously, and so continuously in the Near East and associated areas, there is no mythic literature in these areas associated with its spread.

In the Old Testament, hunting, herding, and farming as modes of livelihood co-exist simultaneously, and there is no explanation in terms of the historical process by which they evolved, although definite modes of social and political behaviour are associated with these various livelihoods in mythic elements that provide the series of exclusions by which the criteria of the collectivity of the 'chosen people' is established. Co-existence of settlements that range from mountainous camps, to rural villages, desert oases, and large urban centers are reported in a matter of fact way. The higher skills of civilization, such as banking, and other forms of economic and agricultural management skills, architecture, medicine, pharmacy, and special forms of magic and psychiatry were seen to be readily available directly from the Egyptian empire as a matter of convenience, as can be seen for example in the myth of Joseph. Other specialized skills such as ship-building and more sophisticated forms of herding and breeding of animals were received directly by the moral man from God, who represented at one and the same time Pharaonic power and cosmic consciousness, as can be seen in the myth of Noah.

However, the early Chinese myths that do exist are concerned with the dissemination of all the skills of civilization by human individuals who contained in their persons great knowledge about these skills and went among the masses to teach them. Legendary figures appear about the time of the Shang dynasty or earlier who taught the population the skills of cooking, architecture, medicine, pharmacology, agriculture, clothing manufacture, engineering and city planning. These myths were set down in a logos of Chinese documents collated only after the fall of the Shang dynasty about 5-6000 years ago, and which in its extant and final form can be traced to only about 2500-3000 years ago. As Chinese civilization had by this time developed a system of historiography which is similar in its approach and methodology to the Greek system of historiography adapted by civilizations of the 4th efflorescence mode, this body of collective culture closely distinguishes between history and myth or history and mythic history. The legendary figures who disseminated the skills of civilization in the neolithic mode are documented by the Chinese themselves as the early myths, and not the history, of their civilization.

Thus, the process by which Chinese civilization developed was extremely unique. Removed as the Chinese were from the mainstream of the aftermath of the Egyptian civilization in the Near East, neolithic civilization developed independently and in isolation from the development of neolithic civilization in the rest of the world. Almost immediately after—at least by the standards of historical time—the development and dissemination of neolithic civilization on the Chinese mainland there occurred, both independently and in isolation from the rest of the world, a rationalization of culture that resembled the Greek model, where there was a distinct if not an absolute separation between subjectification and objectification of phenomena.

The Chinese were able to separate myth and history, religion and politics, magic and science in a way that the early civilizations of the Near East were unable to do. In the Near East proper this distinction did not become established in fact until the emergence of Islam as a religious philosophy some 3-4000 years after this distinction had been made by Chinese civilization. In fact, remote from Greek civilization by geography and historical circumstances, this rationalization of Chinese culture occurred at about the same time or earlier as the establishment of rational culture by the Greeks in the time of Socrates about 2500 years ago. Rationalization of culture by the Chinese however cannot be confused with the objectification of phenomena mode of the Greeks, for the Chinese, in their rationalization of culture, made a clear if not absolute distinction between subjectification and objectification of phenomena, while the Greeks in their rationalization of culture simply denied subjectification of phenomena as a valid mode of conceptualization and communication. This mode was different again from the Indian which fused subjectification of phenomena modes into objectification of phenomena modes, and in this way found it an irrelevant exercise to attempt to make any distinction between them.

A critical factor in the unique ontology of Chinese culture and the way in which it was differentiated from all other great civilizations was the form of written language which the Chinese possessed. The Chinese in their early settlements had carried with them a form of written language which its earliest remaining vestiges suggest was entirely similar to the hieroglyphic language of the ancient Egyptians. Removed as the Chinese were geographically from the Egyptian empire and the Near East, the Chinese had no opportunity to acquire those developments in written language that occurred during the aftermath of the Egyptian empire in the neolithic civilizations of contiguous areas. Over a period of 10,000 years or more in areas contiguous to the Egyptian empire and in constant revolt against it, the Babylonian, Sumerian and eventually Semitic and Phoenician written language systems evolved as a matter of a particular historical process. The original Egyptian writing system was developed by the political elite of the enslaved masses of the Egyptian empire as they organized the knowledge of the peak period of 3rd efflorescence society into civilization skills.

These skills were organized for the particular purposes of the Pharaonic elite, and as such were directed in terms of the technological, magical and religious systems which were meant to both ensure the continued domination of the Pharaonic elite over the enslaved masses and at the same time to effect their chances of physical reincarnation. This writing system was basically a system of signs or codes, rather than a language effective for purposes of ordinary everyday communication. From the point of view of the political elite of the enslaved masses this sign and code system referred to and keyed into the vast areas of knowledge which they possessed in various areas and on different phenomenological levels. From the point of view of the Pharaonic elite these signs and codes represented a means of objectifying and verifying the purposes for which the writing system had been developed.

The Pharaonic elite was laregly ignorant of the details of knowledge to which these signs and codes referred, and used them only as checks in tabulating and judging the effectiveness of the specific and visible purposes to which this knowledge was used. In the early stages of the original Egyptian empire it was likely that the enslaved masses possessed only a rudimentary form of spoken language, where visible and recognizable words referred to vast stores of knowledge conceptualized in the modes of subjectification of phenomena. Spoken language was extremely contextual, where the same word could refer to entirely different levels of phenomena, or areas of common knowledge, depending on the context, which was defined by a small set of linguistic markers or particles.

Organization of the knowledge of the peoples in a written language for specific purposes meant that the contextual relativity of the language decreased, as written symbols accumulated designating different areas of knowledge and different levels of phenomena. As knowledge became organized, and over time the skills of civilization more evenly disseminated among the masses due to the large scale social, economic and technological needs of the empire, language became increasingly less relative and less contextual, more and more symbols accumulating to refer to specific areas of knowledge and particulars within these areas of knowledge. Language became a mode of communication about specific skills of civilization and their particulars among large and often dispersed groups who engaged in a systematic and continuous exchange of these skills and their particulars. Whereas earlier forms of language had been limited in the number of different symbols, due to the almost entirely subjective conceptualization of phenomena, and where skills of civilization were general rather than particularized, later forms of language demanded an increasingly larger number of different symbols, where skills of civilization became specific and particularized. This was the context of neolithic civilization both within and without the Egyptian empire in which modern language forms were developed, beginning with the Babylonian cuniform as an intermediary between Egyptian hieroglyphics and the phoneticized Semitic and Phoenician languages.

For the purposes of the Pharaonic elite, the Egyptian hieroglyphic system was sufficient as a mode of written communication and perhaps resembled the spoken language modes with which the Pharaonic elite continued to communicate with the political elite of the enslaved masses. The purposes of the Pharaonic elite were never to exchange particularized skills of civilization, but were single-mindedly bent on maintaining the technical, magical and religious systems that were critical to the entrenchment of their own modes of ruling the empire. For the masses, however, as neolithic civilization developed under the pressures of the social, political, economic, and technological needs of the empire, this language form, both written and spoken, was insufficient. The symbols of the Egyptian hieroglyphics which were used for purposes of code and signifcation were shortened and simplified to emulate in a visual form the actual spoken languages of the masses.

Although information, even in neolithic civilization was more particularized and abundant than in earlier times, it was nonetheless still both limited and specific, and hence the number of symbols used, although larger than in earlier periods, was also both limited and specific. By shortening and simplifying the symbols of Egyptian hieroglyphics for the needs of the masses, these shortened and simplified symbols no longer represented whole areas of knowledge or multi-leveled phenomena. They no longer represented repetition of ideas in different areas of knowledge or multi-leveled phenomena in relationship to some 'master code' that signified methods of conceptualization in terms of the subjectification of phenomena. This was not possible because, for one thing, the masses did not possess in any holistic way this 'master code,' which, in its difficulty and sophistication, remained in the possession of the political elite of the enslaved masses and the Pharaohs, and which continued to be expressed in hieroglyphic form. It was also not useful for the masses, as it prevented easy and quick exchange of the specific skills of civilization and its particulars in everyday communication. Repetition of ideas was confined to those areas that were particularly important in such exchanges, and ideas worthy of repetition were symbolized, representing such things as time, space, action, substance, intent, and person.

As the written language system was developed as an adjunct to the developing spoken language of the masses in neolithic civilization, the symbols were given strong phonetic associations in order to facilitate the relationship between the spoken and the written language, where the spoken language was the primary mode of communication, and the written language, only the secondary one. The original Egyptian hieroglyphics was the means of organizing knowledge towards specific purposes that still existed almost entirely as a subjectification of phenomena mode. As such it was systemic and formulaic, and did not have the purpose of communication in the specifics of various fields of knowledge and their particulars. While it represented a highly organized mode of thought and communication, it did not have the purposes of everyday linguistic communication. Those repetitions of symbols to include such things as time, space, substance, action, intent and person embody the logos of grammar that is common to modern Near Eastern languages and their extreme evolutionary forms in languages known as 'Indo-European.'

In these Indo-European languages derived from neolithic languages at the tail end of the period of neolithic civilization, such as Pheonician, and of which Greek is the prototype, the original references of grammatical components to the 'master code' which organized the wealth of knowledge in subjectification of phenomena modes were all but forgotten. These grammatical particles were accepted as a matter of practical utility without any concern about that to which they referred. The phonetic particles of the language which were also simplified and shortened forms of symbols that had once referred to aspects of the 'master code' were also accepted merely as a matter of utility in terms of their purely mnemonic function. The dissemination of such simplified and shortened symbols for both grammatical and phonetic purposes was undoubtedly the result of the fact that in the Egyptian empire many diverse groups, in many diverse stages of efflorescence, lived together or contiguously. In an effort to exchange and communicate more specific skills of civilization and their particulars which were at first necessary to the needs of the Egyptian empire, and later even more necessary to revolt and independence from it, written and spoken languages conveniently regularized and transmitted by a certain limited number of grammatical and phonetic components easily spread and became accepted as a matter of practical utility.

In the East Asian mainland however, the original colony of the ancient Egyptian empire established some 10-15,000 years ago, like an abandoned child, was forgotten and neglected during the long decline of the mother empire so far away and inaccessible. Stabilized and entirely controlled by its own surrogate Pharaonic rulers and the designated political elite of the enslaved masses there, the ancestors of Chinese civilization had little or no opportunity to be inspired by the revolutions of others, and removed as they were from the hurly burly of the empire's center, had little or no opportunity to engage in an exchange of the specialized skills of civilization and their particulars under the immediate demands of empire. Presumably sent or exiled to the East Asian mainland as a relatively homogeneous group, the homogeneity of the prototypical Chinese became even more homogeneous over time, having little contact with the influx and variations of the races that ebbed and flowed in the Near Eastern regions contiguous to the Egyptian empire.

As the Egyptian empire declined and this far flung colony became more and more remote to the immediate needs of the empire in the throes of suppressing incessant revolutions and rebellions in or near its own territory, there was a halt to communication with the mother empire, and flow of immigration to the East Asian mainland from the Near East most certainly ceased. As neolithic culture developed in the Near East, and the independent kingdoms of the Babylonians, Sumerians and Hebrews and others emerged and flourished, the language systems of these peoples who wished to erase the memory of the origins of their own civilization in the Egyptian empire anyhow, removed them further and further away from the body of knowledge and information that would have made them aware of similar derivative civilization lines in other parts of the world.

Nonetheless in the encyclopaedic logos of Greek mythology, it appears that the Greeks were at least aware of the existence of civilization in China. The reference to the 'Cimerians,' who were alleged to be people of mystery on the farthermost bank of the peaceful river, 'Ocean,' that surrounded the world, suggests the Greeks were referring to China and to Chinese civilization. These people were characterized as obscure, their land was covered by clouds, the sun never shone on them, and they lived in darkness. These mythological allusions suggest that the river of great peace was the calm Pacific ocean, and the general obscurity attributed to these peoples is no doubt a reference on one level to the general lack of contact and accessibility that the Greeks, or those Near Eastern peoples from whom they culled this information, enjoyed with them.

The interesting reference to the fact that the sun never shone on the 'Cimerians,' either by day or by night, is probably an allusion to the particular religious philosophies of the Chinese, particularly as they had begun to develop by about 5000 years ago, at about the same time as the Greeks learned of their existence. The religion of the ancient Egyptians held that the sun, as an object of the natural world, embodied the power and force of the Pharaonic elite through the worship of which the entire collective of the enslaved masses received all the benefits of both nature and society. The living divinity 'Ra' was associated closely with the sun. In the 'master code' of 3rd efflorescence, where knowledge was organized for purposes of 2nd and 4th efflorescence modes of the Pharaonic elite, the 'sun' was one of many symbols that contained both general and specific meanings on many phenomenological levels. Among these other natural and astronomical symbols that were used in the 'master code' of Egyptian knowledge, the Greeks had blithely picked up most of them and associated an individual God, in humanized form, with each of them. Thus, while in Greek mythology, Apollo was the God of the sun, Diana, the Goddess of the moon, and Mars was associated with the planet Mars and so forth. In the almost total objectification of phenomena mode that was characteristic of Greek mythology and philosophy, the original synthesis of all Egyptian knowledge in a 'master code' that depended largely on subjectification of phenomena mode was ignored. In China where rationalization rather than objectification of subjectified phenomena was achieved, however, this 'master code' was kept, although reformed in later periods. This 'master code' remains extant today in the Chinese *Book of Changes* or *I-Ching*, which in its reformed version continued to excercise enormous influence on all of the religious philosophies in China as they evolved.

Of particular importance in these religious philosophies is the symbol of the moon, which is always favorably compared in Chinese religious philosophies to the sun. In Taoist texts it is the 'yin' or moon which is worshipped, and which provides the method by which the 'way' or 'Tao' can be achieved. The 'yin,' from the point of view of modern psychological theories represents the rational consciousness, while 'Tao' represents cosmic mind or consciousness. The moon or 'yin' is always identified with the female, while the sun or 'yang' is always identified with the male. As female ego, 'yin' represents the way by which ego is negated through rational consciousness in order to attain 'Tao' or cosmic consciousness. 'Yang' or sun on the other hand represents the opposite and most pejorative ego in its full expression of libido aggression. Thus, this preoccupation with the 'moon' as the chief symbol of religious philosophy of the Chinese, as opposed to the sun, which was current in all Chinese literature and thinking by about 5000 years ago when the Greeks might have either directly or indirectly encountered it, accounts for the way in which the 'Cimerians,' apparently the Chinese, were characterized in Greek mythology, i.e., as people who were characterized by the absence of the 'sun' in their land, and as people whose depth of knowledge came directly from the 'moon.' As clouds are also often identified with 'yin' in Chinese Taoist texts of this period, the fact that the land of the 'Cimerians' was covered by clouds is another allusion to the Chinese religious philosophy of this time.

In addition although there are general obscure references in the Old Testament to lands or cities of great mystery far to the east of the regions of the Near East, such as Ur, the land from which Abraham was exiled, it is unclear as to whether or not this is simply another covert reference to the ancient Egyptian empire itself or one of its earlier derivative empires in the Near East, or was actually a reference to a distant, highly developed civilization to the east of all of these places in China. By the time the New Testament was written 2000 years ago the Greeks had long been aware of the 'Cimerians,' probably for several thousand years, so that the allusion to the 'wise men' whom Herod used to predict the time and place of the Saviour's birth by astrological and astronomical methods no longer current in the Near East at that time, suggests that these 'wise men' came from a civilization, perhaps China, where the science and technologies of Egypt were still familiar enough to be practiced, at least in their less offensive forms. The most telling example however in the Hebraic Bible that suggests that the existence of China was known to the authors of the Old Testament, is the reference in the Book of Genesis to the 'tree of life.' Adam and Eve were forbidden to eat, both of the 'tree of knowledge' and the 'tree of life,' and after Adam and Eve disobey God's commandments and eat of the 'tree of knowledge,' God becomes fearful that they will then eat of the 'tree of life.' It is this fear, not their original sin, which induces God to banish Adam and Eve from the Garden of Eden. In Genesis, Chapter 3, after Adam and Eve have eaten of the tree of knowledge and been metted the various punishments of mortality and sexual conflict, the narrative explicates:

> And the Lord God said, Behold, the man is become as one of us, to know good and evil: and now, lest he put forth his hand and take also of the tree of life, and eat and live for ever: Therefore the Lord God sent him forth from the Garden of Eden, to till the ground from when he was taken. So he drove out the man; and he placed at the East of the Garden of Eden Cher-u-bims, and a flaming sword which turned every way, to keep the way of the tree of life.

As most of the myths of the Old Testament can be interpreted on many levels of phenomena at the same time, God here represents, on a religious level, cosmic mind and, on the political and social level, the Pharaonic elite. The taboo against the tree of knowledge, from an evolutionary perspective, represents the acquisition of ego-recognition which from both a religious and psychological and cultural point of view bode great trauma for the future of mankind. From the political and social point of view, the forbidden act of taking the fruit from the tree of knowledge, eating it, and then consequently being punished for it, suggests an act of theft on the part of either or both the political elite of the enslaved masses or the enslaved masses themselves, that involved some part of the exclusive knowledge of the Pharaohs. While Adam and Eve, however successfully, stole the fruit of the tree of knowledge, they were never trusted thereafter and deliberately prevented from having access to the tree of life, which was set by God in the east of the Garden where it was protected by various formidable obstacles. As the fruit of the tree of knowledge represented the acquisition of knowledge about self-conscious mortality and ego, and the tree of life represented cosmic libido and the knowledge that would lead to immortality, or at least an extension of human life, it stands to reason that on the social and political level, the theft of knowledge from the Pharaonic elite had only been partially successful. The fruit from the tree of knowledge represents specific skills of civilization that centered around psychology, social organization and economics, while the tree of life represented those skills of civilization associated with what we today regard as the 'hard' sciences, of medicine, engineering, physics, chemistry, astronomy, and so forth.

As the tree of life could lead to a knowledge of immortality or at least an extension of human life, while the tree of knowledge could not, the tree of life, as representing a body of knowledge possessed by the Pharaonic elite, was seen as the greater and more valuable part of that body of knowledge. Since the fruit of the tree of knowledge had endowed Adam and Eve with rational consciousness, ego-recognition, and superego, where cosmic consciousness existed prior to the eating of the fruit, the tree of life did not represent, on a psychological level, any of the major components of individual psyche, nor on the evolutionary level, any of the stages of efflorescence from 1-4, as they were all embodied in the tree of knowledge, although the tree of life did represent cosmic libido and also perhaps collective libido. As a holistic symbol, the tree of life with its scientific associations and intimations of immortality, suggests that that which Adam and Eve failed to attain by theft from the Pharaonic elite was the 'master code' of the wealth of knowledge that the Pharaohs possessed in scientific form. The fact that the allusion maintaining that the tree of life was put in a safe place where it was guarded and where those people of the Near East could not obtain access to it, in the Far East, due to a variety of obstacles, suggests that the authors of the Old Testament were at least aware of the Egyptian colony in China where the 'master code' was known and was, from the point of view of the Pharaohs, in safe-keeping. In addition, 'Cher-u-bims' suggests a possible transliteration of the Greek 'Cimerians.'

The very subject of the Old Testament, on a political and social level, is the continuing rebellion and revolution against the Pharaonic elite in Egypt both within and without the empire in contiguous regions of the Near East. Genesis documents in the myth of Adam and Eve only the first of many such rebellions and consequent exoduses from the empire by successful rebels against the empire of which the myth of Moses later in the Old Testament is the most precise, most historically rather than mythically oriented, documentation of such rebellions and exoduses. At the very early time of these first rebellions which may have been 10-15,000 years ago, the Egyptian colony in China had just been settled, and was not only in contact still with the empire but also under its firm control. Removed by geographical barriers from the centers of rebellion in the empire itself, the Egyptian colony in China at that time, from the point of view of the Pharaohs, would have seemed inviolable and a prized possession indeed. In the course of thousands of years, however, as rebellions and revolutions became more and more frequent and successful neolithic civilizations developed and spread throughout the Near Eastern regions, this distant colony, isolated and remote, became irrelevant to both the Egyptian empire and the noelithic civilizations which grew up to politically and socially displace it.

Concomitant with the political and social rebellions and revolutions against the Egyptian empire were the pejorative attitudes of the successful rebels and revolutionaries towards Egyptian religions and the systems of magic and science associated with them. As the special skills of civilization and their particulars were spread and disseminated in the Near East, with the accompanying simplified and regularized language forms that developed to facilitate their exchange and communication, the original body of knowledge, the 'tree of life' from which these skills had been originally derived, was no longer relevant. It did not serve the purposes of the later stages of the 3rd efflorescence society, with vestiges of 4th efflorescence, that was emerging in these regions, for the 'master code' or 'tree of life' had been the purview of the very few, the Pharaonic rulers and certain of their loyal, politically designated elite from the enslaved masses, who no longer existed in neolithic civilization. The body of knowledge of peak 3rd efflorescence society organized in a 'master code' had served the purposes of 2nd efflorescence modes possessed only by the Pharaonic elite, and it was the purpose of this 'master code,' rather than its substance, that resulted in its disappearance from the Near East and other contiguous regions.

In these areas it was only the Greeks with their intense curiosity due to their objectification of phenomena modes, who eventually revived science as a legitimate and serious subject for study and investigation. By this time however some 2500 years ago, the scientific and technological legacy of the Egyptian empire existed in the Near East and contiguous regions only in bits and pieces, and it is fair to say that the entire history of scientific development since that time in 4th efflorescence societies has been an attempt to piece together, through application of rational consciousness, the fragments of science and technology left by the remains of the Egyptian legacy and form them to purposes compatible with 4th, rather than 2nd or 3rd efflorescence modes.

However, as neolithic civilization began to emerge independently among the Chinese people, the simplified grammatical and phonetic language systems of the Near East were not available to Chinese as they developed the need to exchange and communicate the specific skills and their particulars characteristic of neolithic civilization. The limited variables which came to characterize Semitic and Indo-European languages were drawn from the 'master code' of Egyptian knowledge, and in their grammatical modes depended on the mixture of metaphysical, scientific and psychological concepts that had been originally synthesized in the 'master code.' Space and time in grammatical form were drawn from concepts that were metaphysical and scientific, while person reflected the psychological concepts embodied in the 'master code.' The series of changes in grammatical form, involving conjugation and declension, were a reflection of the precise logical and mathematical concepts that characterized this code in its scientific and metaphysical aspects.

When the neolithic conditions arose in China that demanded a simplified language form in which special skills of civilization and their particulars could be communicated and exchanged, the historical process which was responsible for the development of Semitic and Indo-European languages was unknown and unavailable. The evolution of Semitic and Indo-European languages had happened as a matter of historical accident and reflected the historical necessities of conditions in the Near East as the Egyptian empire fell into decline. Those peoples who developed and adapted the Semitic and Indo-European language forms, over a period of time, became less and less aware of the sources from which their language developed, and accepted their language forms, like neolithic culture itself, as a matter of practical necessity without any undue introspection. The religious philosophies of the Near East, and the rational philosophies derived from them by the Greeks, were seen as a *de facto* explanation of the cosmos, without any necessary relationship to the development of the practical necessities of life in neolithic civilization and without any necessary relationship to the languages in which these practical necessities were carried out in communication and exchange.

The situation in China was very different, as neolithic civilization under existing conditions appeared clearly to be a deliberate, man-made development. The very same political elite who had carried out the will of the Pharaonic surrogates in China were those responsible for the development and dissemination of neolithic civilization there. As the collectives in China were almost entirely racially homogeneous, the political elite of the enslaved masses, the Pharaonic surrogates, and the enslaved masses themselves were all members of a common collective, unlike the Near East. Due to formidable geographical barriers, influx of Nordic hordes from the north, northeast and northwest was rare, but when it did occur. unlike the situation in the early periods of the Nile Valley, these Nordic invaders did not encounter collectives in the peak of 3rd efflorescence society. Instead they encountered a rigidly structured society along the lines of the Egyptian empire, where the surrogate Pharaonic elite possessed all the attributes of 2nd and 4th efflorescence society and who possessed in addition the wealth of knowledge culled from 3rd efflorescence society applied to the aims of the surrogate Pharaonic elite. Rather than being in a position to conquer these Chinese kingdoms, Nordic invaders from the north, northeast and northwest, were absorbed by them.

However, as the Chinese colony or colonies over time lost contact with the Egyptian empire, the original purpose of a surrogate Pharaonic elite became somewhat obscured on the Chinese mainland, for unlike the situation in the middle east where the Pharaonic elite had derived from collectives with markedly different efflorescence modes than the people they conquered, this was not the case in China where the surrogate Pharaonic elite and the masses they controlled shared common collective modes. The surrogate Pharaonic elite of China had been no more than politically designated elites of the same masses from which they themselves derived, and in whom common modes of collective behaviour were indistinguishable from those over whom they ruled in surrogate fashion, on behalf of the true Pharaonic elite of the Egyptian empire. As contact with the Egyptian empire lessened and eventually disappeared over time, an awareness of the commonalty of the elite and the masses, among both of these classes, began to emerge. There was never, thus, the sense of confrontation and hostility between the ruling elite and the masses of China, which was one of the critical differentiating factors between Chinese and Near Eastern civilizations and their derivatives, and which was responsible for the kind of historical evolutionary development which occurred in China and made it appear, in its refined form, so totally different from the Near Eastern civilizations and their derivatives in India and Greece.

Modes of the 2nd and 4th efflorescence which had been so deeply entrenched in the Pharaonic elite of Egypt were imposed on and learned by the initial Pharaonic elite in China, and as contact with the Egyptian empire disappeared, the natural entrenched modes of 3rd efflorescence society, even among the surrogate Pharaonic elite in China, emerged in a more dominant form. Entrenchment of superego as a mode of collective behaviour began to supercede the superficial 2nd and 4th efflorescence forms of this elite, and the organized purposes to which peak 3rd efflorescence knowledge had been directed became more diffuse. Thus, as early Chinese myths explain, it was the ruling elite who, as a matter of their beneficent attitude towards the masses they controlled, were responsible for the creation and dissemination of the special skills of civilization and their particulars which characterize neolithic civilization. In an entirely self-conscious way, the political elite of China took those aspects of the organized knowledge of the Egyptian empire which were useful for the welfare of the masses and deliberately disseminated it. The skills of neolithic civilization were not acquired by theft or revolution as in the Near East, but were voluntarily given by the political elite to the masses with whom they shared basic and fundamental collective modes and behaviour and with whom they consequently felt a great empathy. Only after this process had enabled neolithic civilization to become entrenched among the Chinese masses, and when, in certain periods a surrogate Pharaonic elite once again attempted to establish total control over the masses, did revolutions begin to occur in China.

Revolution against a Pharaonic elite, legitimate or surrogate, was never possible until the masses had first acquired basic skills of neolithic civilization, and while in the Near East these skills were obtained by theft, subterfuge, and rebellion in reaction to the Pharaonic elite, in China these skills were voluntarily given to the masses by the surrogate Pharaonic elite. There are no corresponding myths in China to Adam and Eve eating of the fruit of the tree of knowledge in contravention of the law of God, nor any corresponding myths to the Greek myth of Prometheus who stole the fire from the Gods for man in contravention to their laws. The denouement of these myths where Adam and Eve are punished by expulsion from the Garden of Eden and forced to undergo extreme hardship and conflict, or where Prometheus is punished for his sins by physical brutalization have no counterpart in Chinese myth or legend. Those legendary figures from the distant past some 5-10,000 years ago who voluntarily taught the various skills of civilization to the Chinese masses are venerated in myth and legend, have no pejorative associations and suffer no punishment for their acts.

To Fu-Hsi who, according to Chinese legend was one of these mythic dispensers of civilization and who, according to these legends lived from 5-7000 years ago, is attributed the earliest version of the Chinese *I-Ching* or *Book of Changes*. In addition, the invention of Chinese language in its present form is often attributed to Fu-Hsi. Under the comparatively tranquil circumstances in China where the special skills of civilization and their particulars that comprise neolithic civilization were developed and taught to the masses, these skills and their particulars are seen as having been developed as a matter of deliberation, and not of accident as in the Near East. The sense that neolithic skills were evolved and disseminated as a result of deliberation, rather than historical accident, reflects these comparatively tranquil circumstances under which neolithic civilization was disseminated in China. In fact, under the demands of the Pharaonic type kingdoms in China, both political, social and economic, the spread of neolithic civilization with its needs for the exchange and communication of special civilization skills and their particulars was no doubt just as much a matter of historical process and necessity as in the Near East.

The relatively conflict-free aspect of these processes, however, led to the sense that the spread of this civilization was deliberate rather than accidental, and indeed, given the common collective and racial modes of the elite and the masses, dissemination of these skills was no doubt facilitated by the attitude of the elite which differed so markedly from the attitude of elite groups in the Near East whose collective and racial modes were distinctively different from the masses under their control. Thus the mythic modes in which the dissemination of neolithic civilization is expressed in China, although to some extent no doubt an exaggeration of the actual facts, does reflect certain aspects of this process, just as the mythic modes in which the dissemination of neolithic civlization in the Near East is also an exaggeration of the facts reflecting certain aspects of the actual, conflict-ridden process by which such dissemination was achieved. The relationship between the elites and the masses in these various societies, Near Eastern, Hebraic, Indian, Greek and Chinese, however, as they were conceptualized in early myths and passed down from generation to generation as aspects of cultural and collective modes of behaviour have been critical factors in determining the evolutionary end of these various civilizations and the marked differences among them in their attitude towards entering into 4th efflorescence society or remaining in the late stages of the 3rd.

The attitude towards elites in the Near Eastern and Hebraic myths is conflict ridden and often confused, and elites are seen to be acceptable only under a set of specific exclusionary rules that are often religious and economic. In Indian myths, social and political elites are regarded pejoratively, but elites as spiritual leaders and guides receive the highest social sanctions as long as they renounce elitism in purely social and political spheres. In Greek myths, elites are characterized by their aggressive characteristics, but these characteristics are deified as the sanctioned mode of behaviour which is to be emulated by all members of the collective in social and political behaviour. In Chinese myths, elites are neither aggressive nor embody modes of social and political conflict. Their substantive elitism is embodied in their patronization and beneficence towards the masses. They are not seen as spirital leaders *per se* but as exemplary models of moral behaviour, and their modes of patronization and beneficence are seen as worthy of emulation by all members of the collective who have not, largely for practical reasons, been able to attain the special moral, technical and social skills which differentiate them from the masses and make them truly elite. Although all these myths and the peoples who possess them have a common historical thread in their ultimate derivation from the Egyptian empire, the variable historical circumstances in which they each developed account for the final configuration of the attitudes they possess towards elites. As attitude towards elites determine the ultimate ego-recognition and expression of libido aggression of the collective, the transformation of 3rd efflorescence modes into 4th efflorescence modes, these various attitudes towards elites among these various civilizations determine the ultimate potentiality of these collectives in 4th efflorescence modes.

In fact, the Greek model, where elites are sanctioned in their aggressive characteristics, or where the ego-recognition and expression of libido aggression of the collective is encouraged and sanctioned, is the only model which has come full-fledged into 4th efflorescence society in our times. Where elites are not fully sanctioned in their ego-recognition and expression of libido aggression as in Near Eastern civilization, or where these modes are transformed into cosmic, rather than collective levels, as in Indian civilization, these societies remain in the late stages of the 3rd efflorescence. In China where, although elites are sanctioned, they lack attributes of ego-recognition and expression of libido aggression, the society also remains in the late stages of the 3rd efflorescence. This is not to say that there may not be other modes besides the Greek that may in the future enter into 4th efflorescence society. In societies of the 'under-developed' world in the late stages of the 3rd efflorescence, however, the importance attached to ego-negating and libido abdicating elites may yield the potential for societies that may enter into the 4th efflorescence at the same time as they possess modes of a new, 5th efflorescence society, where the ego-recognition and libido aggression of the collective self may be mitigated by a cosmic, as opposed to merely collective, ego-recognition and libido aggression.

Whether or not humanoids as a species or a genus will be able to survive at all may depend on how well and how quickly they are able to adjudicate rational consciousness with cosmic consciousness in the collective ego-recognition and expression of libido aggression in 4th efflorescence society. For such survival to be ensured, it will be necessary to call on the collective experience of all human civilizations as they have evolved and as they have developed different attitudes towards the relationship between the elite and the masses. It will also be necessary to be able to differentiate among these collective experiences as to what is universal and common to the needs of the species as a whole and what is the result of particular historical experiences whose lessons are obsolete and no longer relevant. For this subject, the evolution of Chinese civilization is of particular interest, since, unlike most other civilizations developing out of the common historical thread which has its origins in the Egyptian empire, its evolution was a matter of, comparatively speaking at least, slow deliberation subject to, to a rather unusual extent, the application of rational consciousness and introspection in many phases of its development.

This is not to suggest that Chinese civilization is superior to other civilizations with whom it has a common thread, but only to suggest that in the way that Chinese civilization developed, which was in the first instance a matter of historical accident, there may be important lessons to be learned. Certain features of Chinese civilization allow us to appreciate that the evolutionary process may be regarded as a self-conscious process, the result of the application of rational consciousness, and thus subject to a teleology which is within the human ken. Above all, a teleology which is within the human ken allows for the survival of the human species or genus, and all forces which would mitigate against that survival, whether individual, collective or cosmic, must be treated in a pejorative way. Thus, these forces can be shaped into modes that are more favorable in terms of the ultimate advantageous adaptions they bode for human survival.

Chapter XIV

THE CHINESE LANGUAGE SYSTEM: LANGUAGE AS ABSTRACT ART

The self-conscious development of a simplified written language that could be used for the exchange and communication of special skills of civilization and their particulars in neolithic culture in China is personified in the legendary figure of Fu-Hsi, who, according to Chinese legend, lived 5-7000 years ago. In the long periods of time between peak 3rd efflorescence society and the late stages of the 3rd or the 4th, when there were no simplified forms of written or even spoken language, the complexities of historical processes were condensed in mythic forms. It is one of the enduring characteristics of Chinese civilization that the processes of historical evolution tend to be expressed in personae of extraordinary talent and whose morally exemplary lives are the subject for continued veneration even into modern times. Because the Greeks entered onto the stage of neolithic civilization near its end and as strangers to the direct experience of its development, the legendary figures of Greek myths represent the objectification of the experiences of other collectives and are treated in a rather matter of fact way, neither particularly subjects of either veneration or denigration.

It was in fact this lack of direct experience with these legendary figures that led to the general breakdown of Greek religion about 2500 years ago, when history and philosophy were developed among the Greeks who, by this time, possessed an objectified language system that could express with great precision their almost total objectification of phenomena modes of thought and conceptualization. It was this mode of objectification of phenomena which belonged to the Greek collective as a matter of direct and intimate historical experience, while the legendary figures who comprised the logos of Greek religion were sensed as inauthentic in terms of the direct experience of the collective and hence easily dispensed with. Analytical history and philosophy replaced myth as a medium of historical expression among the Greeks not only as a matter of evolutionary development, but also as a matter of authenticity in terms of the direct experience of the Greek collective.

Among the Near Eastern peoples legendary figures and mythic parables of their civilization were retained because they reflected the direct and authentic experience of the collective in its historical evolution. Despite increasing objectification of these societies into modern times, these legends and mythic elements were maintained as they set the parameters of the religious philosophies of ego-negation and libido abdication which marked their ultimate evolution into societies in the late stages of the 3rd efflorescence. Not until the time of Jesus Christ 2000 years ago did the Greek collectives in their derivative civilizations in southern and central Europe find a legendary figure who could express the direct experience of these collectives in an authentic way. By the time of Jesus, objectification of phenomena modes had become an accepted and integral part of collective modes of Greek societies and its derivatives and represented *de facto* the logos of its historical experience.

The persona of Jesus Christ allowed the Greek societies and their derivatives to synthesize their modes of objectification with the deeper modes of subjectification of phenomena from which objectification of phenomena modes had been originally derived, and which had come to seem irrelevant to Greek societies and their derivatives in the direct experience of their own historical evolution. The persona of Jesus Christ allowed the Greeks to adjudicate rational consciousness with cosmic consciousness. For Greek societies and their derivatives direct experience with cosmic consciousness was a new experience which occurred at the point where all the modes of objectification of phenomena had been carried to their logical extreme, and found wanting. Jesus Christ as a persona represented not only the morally exemplary life but also the direct experience of a collective with all the vicissitudes of objectification of phenomena modes which, in the end, were found to be inadequate as a way of explaining or relating to reality, individual, collective and cosmic.

Whereas the persona of Jesus Christ made a clear distinction between phenomenological levels of the individual, the collective and the cosmic, the Indian Buddha or Buddhas did not. In India objectification of phenomena modes, as they encroached from the Nordic Aryans and emerged from the late stages of neolithic civilization among the southern Vedic peoples, were entirely absorbed in the service of a subjectification of phenomena mode embodied in the cosmic mind or cosmic consciousness. Rational consciousness as a collective mode never developed in Indian society as an independent psychic component and where it arose it was subjugated and fused with cosmic consciousness. However, in all of these societies, Near Eastern, Hebraic, Greek and Indian, however they were handled, bifurcation of subjectification and objectification of phenomena existed as a potentiality, due to the development through historical processes of the Semitic and Indo-European language forms. In China, however, no such clear kind of bifurcation of subjectification and objectification of phenomena modes was ever possible, due to the independent manner in which the Chinese language form developed.

While Chinese culture was rationalized, it was never possible to clearly bifurcate individual, collective and cosmic consciousness due to the nature of the developed language form. On the other hand, while rationalization of Chinese culture did not permit this bifurcation, it did enable each level of consciousness, individual, collective and cosmic, to infuse itself on every other level at the same time. In China consciousness was able to develop in a holistic way, and as a continuum from the individual, to the collective and to the cosmic levels. The cumulative effects of such a continuum, where the individual, collective and cosmic levels of consciousness infused themselves on each other, both backwards and forwards, in a series of logical steps that were only minutely distinguishable, was to strengthen consciousness as a whole, and to place it in clear contradistinction to those other psychic drives which have, on the whole, proven more destructive than constructive for the welfare of the individual and the collective to, the ego, the libido and the superego.

Fu Hsi and those other legendary figures who were responsible for the invention of the modern Chinese language forms were forced to confront problems which the other civilizations coming out of the common thread of the Egyptian empire were able to evade by virtue of historical processes so intricate that in the end they were best summed up in terms of historical accident. The inventors of the Chinese language were forced to self-consciously select from the 'master code' of the Pharaonic elite and the symbolic language of hieroglyphics which was its official manifestation, those elements that could be used to regularize and simplify a language form suitable to the needs of neolithic civilization. Since the mnemonic methods by which phonetic type languages had evolved in other civilizations coming out of the common thread of the Egyptian empire were inaccessible to the Chinese, for both historical and geographical reasons, the inventors of Chinese language were forced to accept *prima facie* the material available to them in the 'master code' of the Egyptians and their written hieroglyphics which was entirely of a visual, rather than mnemonic, nature. Thus symbols of Egyptian hieroglyphics were broken down in terms of a limited number of visual representations which expressed basic scientific, logical and philosophical concepts derived from the 'master code.'

These visual representations for the most part, of necessity, used a geometrical vocabulary, where lines that were vertical were conceptually differentiated from lines that were horizontal, and both of these differentiated from lines that were curved either to the left or the right. In building the modern Chinese ideographs, short as opposed to long vertical, horizontal and curved lines were differentiated, and dots as opposed to lines also formed part of the basic building blocks of these ideographs. Lines, curves and dots in the various directions signifying north, east, south, and west were combined and resulted in a vocabulary of completed ideographs. In earlier forms of written Chinese whole and half circles were used, but these were regularized in later Chinese writing by about 2500 years ago to completed and incompleted perfect squares and rectangles.

Language symbols that were based on precise, mathematical geometrical representation expressed the scientific and logical concepts of the 'master code.' However, since figures of mathematical geometry were not used for purposes of mathematics but language and communication, the cumulative effect of these geometric figures in their various combinations and permutations was architectonic rather than mathematical or geometric. In this way aesthetics and mathematics were combined to create language forms. Thus, the symbol □, a complete square that is empty, signified a hole, which by extension came to mean 'mouth.' The complete square with a horizontal line bifurcating it 曰 extended the meaning of a 'hole with something in it' to mean 'mouth with something in it' or simply, 'speak.' A closed rectangular shape bifurcated with a horizontal line 日 meant 'sun' or 'day' by extension. This was a regularization of the earlier shape for 'sun' ☉ which was both a pictorial representation, derived directly from the hieroglyphics, and a scientific or metaphysical representation of 'sun.' The representation for 'earth' was two parallel horizontal lines of increasing size from north to south and bifurcated by a vertical line which terminated on the southern end and was left open at the northern end, 土. This was at the same time both a pictorial representation of 'earth,' and a logical and geometric representation of 'earth,' suggesting solidity, depth, and the direction of both of these. In combination with 'sun' 日, 'earth' 土, and a derivative of 'earth' 寸 which had come to mean 'a measurement of distance or depth,' the symbol for 'time' 時 was built.

The symbol for 'moon' was composed of an incomplete rectangular shape, one side of which was elliptically curved and was bifurcated by two parallel horizontal lines: 月. The elliptically curved side of this open rectangular form signifies both geometrically and pictorially the 'moon,' where the angle of trajectory of the moon is elliptical and where it also appears to visual sight, during certain phases, to be of elliptical half-moon shape. The two parallel horizontal lines which bifurcate the figure suggests the mathematical and geometrical aspect of the various phases of the moon, both in its visual appearance, and in the path of its trajectory around the earth. In combination with a number of other limited geometrical shapes that form the vocabulary of Chinese language 'moon' may represent various kinds of 'times,' 'the night,' 'sacrifice,' or 'clothing,' and when two 'moons' are juxtaposed 朋 the symbol represents 'friendship,' so that we see how, in the course of communication and exchange of the special skills of civilization and their particulars characteristic of neolithic civilization, these symbols which are originally mathematical and geometrical in origin, and architectonic in cumulative effect, may take on the highly extended meanings necessary for modern language systems.

In many cases the cumulative architectonic effect of geometric building blocks of Chinese language symbols supercedes any logical or metaphysical concepts that these shapes express. This is particularly true in the simpler and more commonly used language symbols. Thus the symbol 人 or its equivalent 亻 means 'human being' through its suggestion of upright posture and two-leggedness in the two mirror image elliptical curves in either direction with the suggestion of an emerging torso. In the equivalent 亻 uprightness is suggested by the single vertical stroke juxtaposed by the short elliptical curve to the left joined to the top of the vertical stroke, suggesting a head to the body. As a radical or component figure adjoined to others this is one of the most common components in the Chinese language system. Thus combined with a single perpendicular stroke 丨, and a single elliptical curve to the left 亅, 介 means 'alone and depending the context can mean 'alone' in the sense of both 'great' and 'small.' Each of these single strokes 丨 and 亅 have no distinct meaning components in and of themselves, but are taken in Chinese to signify singularity, oneness or singleness. The fact that the one is curved and the other is not, from an induction of meaning in other, larger geometric meaning units, can be taken to suggest sexuality, where the curved elliptical vertical line is usually associated with the female and the straight vertical line is associated with the male.

We can see this dichotomy already in the difference between 'sun' and 'moon' where 'sun' contains only straight lines while 'moon' contains a vertical elliptical one. From the body of religious and metaphysical thought contained in this 'master code' and common to all civilizations coming out of the common thread of Egyptian civilization 'sun' is always given a male persona, and 'moon' a female persona. On the psychological level, 'sun' symbolizes libido and 'moon,' ego-recognition or consciousness. Thus for 'human being' represented by 人 two elliptical vertical strokes in opposite directions there is a representation in an abstract way of 'ego being reflected' or individual consciousness, and this is the concept so aptly used to represent 'human being.' The cumulative effect of two lines, one female and one male, without any geometric symmetry, placed under the figure representing 'human being' is that of isolation and alienation and thus the meaning 'alone' 介 is derived.

Curved lines proper in the word symbol for 'woman' 女 and their components associated with other word symbols to yield extended meanings suggest superego and the 'female type' ego, both through representation and the geometrical or graphic associations they produce. The two curved lines in opposite directions which form the basis of the word symbol for 'woman' 女 is a mathematical model of two ellipses, and two hyperbols which when represented this way produces the helix form . This represents in associative way the continuing cycle of reproduction and in a metaphysical way represents the cyclical form of creation. With an extended line through the word symbol — signifying status, the female is given ego identity, but the associative extension here is that of limiting the female powers in the single female person and also suggests the control of the superego in the development of the 'female type.' The word symbol for 'son' is 子 where the basis is 了 and suggests one half of the female figure or the 'female type,' its product and its complement. Thus the word symbol 又 derived from the word symbol for 'female' means 'to regulate' and 好 where 子 means 'son' or 'child' has the meaning of 'good.' Here 'mother and son,' the quintessence of a society based on the nurture of the 'female type' is given its clear valuation.

The Chinese word symbol 亼 composed of the 'human being' figure with a horizontal line under it means 'to assemble' where the cumulative effect of these strokes directly suggests 'extension of human beings.' Two symbols for 'human being' together as in 人人 extends its single 'human being' meaning to 'follow or obey,' where a group of human beings in two and in tandem directly suggests this meaning.

The dot figure, which is a short elliptical line falling to the right 丶 is usually added to any figure meaning 'addition,' usually 'parental addition.' Adjoined to the figure for 'human being' 个 it means 'wealth,' suggesting by logical induction that the wealth of a human being is the result of his parents' largess. The language symbol 但 composed of 'human being,' 'sun,' and a single horizontal line under 'sun' means 'single,' 'but,' 'only,' or 'merely,' where the singleness of the human being is underlined in another graphic way allowing it to take on an extended meaning of singularity different from 介 which means 'alone.' Similarly the language symbol 乍 containing a 'human being' with two parallel horizontal lines joined to it on the right means 'humane,' and it is not only the double extension of two, as opposed to one horizontal line, that differentiates it from 亼 'to assemble,' but the position of these lines to the east, rather than to the south, which adds to the architectonic effect of this symbol and enables it to possess quite a different extended meaning from virtually the same geometric components. The parallel nature and the right-handed direction of the two horizontal lines here adjoined to 'human being' gives the sense of extension of self and identification of self, in an equal way, with the group or populace.

When a horizontal line is drawn through a geometrical figure or is placed on its northern extremity it usually signifies some kind of social status, and in common folk culture is held to do so by the fact that it resembles in these kinds of configurations, a hat that signifies status. We have seen how a horizontal line drawn through the basic component for female 乂 signifies 'female person' 女, or the individual identity of the female. Thus when a horizontal line is drawn through the figure for human being 大 it means 'great.' When there is an additional horizontal line on top of this figure 天 it means 'heaven,' signifying as it does the ultimate status of God or cosmic consciousness in the metaphysical logos of Chinese philosophy and culture. Thus the horizontal line on the north or top of the figure for earth 土, 王 signifies 'ruler' or 'royalty.' With the added dot that signifies 'addition' or 'parentage' the extended meaning of 玉 is 'jade,' the valuable jewel which is the representation of status or wealth in Chinese civilization.

Thus the single perpendicular line with a slight curve or dot at the end 亅 which is a geometrical radical or component that means 'barb,' topped by the horizontal line signifying status means 丁 'male individual.' This illustrates the status of the male adult as opposed to the female in Chinese culture where the horizontal line in the figure for 'female person' 女 does not close the figure but merely extends through it. Combined with the figure for 'human being' however 仃 the meaning is again extended to 'alone' where the male individual in the context of his humanity and without the mitigating influence of the female is 'alone.'

The horizontal line on top of a figure which gives the impression of full closure through two dots at either end 冖 signifies 'cover' by extended meaning, while the same figure with an additional dot on top by extended meaning 宀 signifies 'roof.' Thus, the symbol 宁 a combination of figures representing 'roof' and 'male individual' means 'the space between the throne room and the retiring room of the prince' or a 'royal hallway.' By further extended meaning the addition of the figure for 'human being' to this word symbol 伫 means 'to stand and wait.'

The radical for 'roof' combined with a figure of two parallel mouths joined on the left by a perpendicular line suggests in a very architectonic way the meaning it holds 官, 'official or mandarin,' where the series of connected mouths express through graphic imagery the role of an 'official' in 'speaking from a series of connected mouths,' that is, bureaucratic talk. Under the protective roof of offical status this bureaucratic, rather than direct speech, is sanctioned. By contrast, a less sarcastic version of the representation of the collective voicing itself through officialdom is given by the word symbol which means 'statesman' 臣 rather than 'official or mandarin' as 官. In the term for 'statesman' there is no protective roof or protected position. Instead a horizontal line is connected to the vertical line on both the north and south termination points of the figure. The closure of these horizontal lines suggests the status of the 'statesman' in a more spare and moral sense than the artificial and technologically created 'roof' suggests. One of the basic units for this figure is 'work' 工 where its graphic representation of status 'heavenward' 天 and solidity 'earthward' 土 are associations of the 'work' component.

These associations place high value and depth on 'work' in a graphic way. In the word symbol for 'statesman' 臣 the components for 'work' are connected both on the top and bottom to the word for 'mouth' 口, yielding a very practical and common sense status for 'statesman' that is lacking in the graphic representation of a series of officially protected mouths in the word symbol for 'official' or 'mandarin' 官. The common sense and practical value of the 'statesman' as opposed to the 'official' can be seen from the extended meaning of 'statesman' which when combined with the figure for 'human being' 卧 means 'to rest.' Thus the natural and practically valued 'working' 'statesman' when put into a purely 'human context' results in 'rest,' for this 'working' 'statesman' is devoid of the ambition and libido aggression which is characteristic of the 'official' as opposed to the 'statesman.' Given a chance to retire from work, he will do so with pleasure and ease, since in fulfillment of the individual consciousness that makes him human he lacks both individual and collective libido aggression, and has no desire to aggrandize these drives through identification with cosmic libido aggression. Thus, within the language itself through concepts derived from mathematics and philosophy, the geometric forms that are Chinese language symbols accrue, through their architectonic effect, an ego-negating and libido-abdicating philosophy which is effected by a strengthening of consciousness on all levels at once, individual, collective and cosmic.

The basic building blocks of Chinese word symbols as geometric forms have mathematical associations that define the precise mathematical knowledge possessed by the ancient Chinese, and which precedes in time the mathematical development in ancient Greece. That this mathematical knowledge was not analyzed and described by the ancient Chinese as it was by the Greeks is a result of the way in which this knowledge was conceptualized and maintained. The mnemonic language forms and long word chains of the Greek language allows this knowledge to be objectified, so that mathematical and scientific thought was the end product of language. In Chinese civilizations mathematical and scientific thought was the method or way of conceptualization, rather than its end product. Thus the negative as word symbol in Chinese such as 不 is the equivalent of a square bisected at right angles and divided into four equilateral right angle triangles. It is a Euclidian type paradigm or geometric model where angles and lines on the perpendicular are cancelled out by angles and lines on the oblique. It is a direct mathematical and geometrical model of negation. Another common negative word symbol 非 represents by mathematical and aesthetic symmetry the sense of perfect mirror image opposition and hence negation. While 不 is an absolute negative, 非 which is composed partly of mathematical, partly of aesthetic, associations means 'no' in the sense of 'wrong' or 'bad.' Thus 非 is combined with many other geometric configurations to produce extended meanings where 不 is not.

Combined with the word symbol for 'heart' or 'mind' 心 or 忄 which in its original form 心 is composed of one curved line to the southeast, with one dot to the west, one dot in the center and one dot to the east, graphically suggesting the 'additions' that comprise 'thought,' the word symbol 悱 means 'desirous of speaking' or 'longing to speak.' The 'heart' or 'mind' wishes to speak but this wish is cancelled by the 'wrong' which prevents it from doing so. When 非 is combined with the word symbol for 'speech' or 'talk' 言 where the 'mouth' 口 is given status by the 'officialness' of the abbreviated 'roof' 宀 from which repetitive extensions and additions of such status is represented by 二, the word symbol 誹 means 'to backbite or slander' as graphically expressed by the juxtaposition of these symbols 'wrong or bad talk with official sanctions.'

Another negative word symbol 無 is formed by associations that are not only geometrical and representative but also arithmetical. The geometric component 皿 composed of four vertical lines means 'utensil' or 'vessel' and encased in an enclosure 皿 is 'enclosed vessel.' The horizontal line extended through it here suggests 無 the breakage of the vessel, leading to a negative sense through representation or graphic symbolism, also suggesting, through mental imagery, the arithmetical mechanisms of subtraction or division. The addition of the 4 dots 灬 which alone means 'fire' further enhances the idea of the 'destruction of the vessel' or a negative. The four vertical lines of the vessel's substance are balanced by the four dots representing fire, where these four dots may also represent the arithmetical remainder of the arithmetical mechanisms suggested by mental imagery. This word symbol has the special extended negative meanings of 'without,' 'apart from,' or 'none,' meanings which in their abstract connotations again suggest the arithmetical imagery involved in the word symbol. This word symbol is also combined with many other geometrical configurations for further extended and specialized meanings.

Combined with the word symbol for 'heart' or 'mind' 心 in 憮 this negative means 'disappointed' or 'disconcerted,' the combinations of word symbols producing the architectonic effect of 'apart from mind or heart,' and hence 'disappointed' or 'disconcerted.' With the word symbol for 'grass' or 'flowers' 艸 or 艹 this negative has the further extended meaning of 'jungle' or 'wasteland,' the architectonic effect of the combination of the juxtoposed word symbols being 'without grass or flowers' and hence 'jungle' or wasteland.' With the symbol for human being 人, the symbolic addition 儛 is 'to skip around' or 'to dance for joy,' where the architectonic effect of the combination is a literal symbolism that means 'without rational consciousness.'

In a language system where all word symbols are composed of geometrical forms that have associations with logical, philosophical, or representative associations, a clear distinction between noun or substantive words and verbs or processual words is impossible. Thus, those geometric forms which have associations that express action, or the lack of it, as in these negatives, are used as verbs. Thus a word form such as 悱 which means literally 'heart wronged' has the verbal meaning 'to desire' in the context of 'desiring to speak one's mind.' The word symbol 誹 which means literally 'wrong speech' has the verbal meaning 'to slander.' The word symbol 憮 which literally means 'apart from the heart' has the verbal meaning of 'to be disappointed,' and the word symbol 儛 which means literally 'man apart' has the verbal meaning of 'to dance for joy' or 'to be elated.'

It can be seen in all cases of Chinese word symbols that each word symbol, and its component meaning units, can never really be translated in any precise way into a language form such as ours which is based on modes of objectification phenomena. Each word symbol is similar to an abstract picture, or a mathematical or geometric model which we grasp immediately through subjectification of phenomena, and only later attempt to put into objectification of phenomena modes. Each word symbol calls forth visceral images which are a composite of both instinctive and cognitive reactions. Through the shape of the symbol we are 'reminded' of a series of objects and not a single object buried in both our unconscious and conscious minds. These symbols arouse not just thought but also feeling. As such they arouse and engage both libido, superego and ego. The cognitive associations of the word symbol at the same time engage and arouse consciousness on all levels and thus allow the amelioration and integration of libido, superego and ego by consciousness simultaneously with the arousal and engagement of these powerful psychic drives.

It can be seen how difficult it is to clearly separate cultural concepts as such in Chinese civilization from the word symbols which represent them. For key concepts of Chinese culture are portrayed by precise conceptual associations of the geometric figures which express them. This is similar to writing about a particular concept in an objectified language form and making a painting of it all at the same time. Only in modern films can we sometimes obtain the same elevated mental effects that the Chinese writing form consistently achieves as a matter of normal, everyday communication in what are usually, from the point of view of objectification, rather mundane matters. Key concepts in Chinese culture thus have a significance in their written expression that key concepts in other civilizations ordinarily lack. In written form, these concepts are not only words or linguistic expressions of the concepts, but approach the substance of the concepts themselves as graphic representations of these concepts. When we realize that these concepts are expressed as words, symbols and signs simultaneously, we begin to understand the significance of the Chinese art of calligraphy. In Chinese calligraphy, whose subject is usually the expression of key concepts of Chinese culture either as single word symbols, or in the guise of a poem or aphorism, the precise movements of the brush, the amount of space and time allocated to each stroke or part of the stroke of the word symbol are able to infuse greater and more subtle depth, as well as the personal involvement and engagement of the calligrapher, into the concept, which is both word, symbol and sign all at the same time. Subtle nuances within each part of the word symbol and between a chain of word symbols can be precisely expressed, according to the cognitive and emotional discretion and judgement of the calligrapher.

Chapter XV

FROM LANGUAGE REFORM TO POLITICAL REVOLUTION IN ANCIENT CHINA

In particular, where word symbols contain action conponents and associations in their geometrical and graphic configurations, the artistry and sensitivity of the calligrapher is permitted an extraordinary degree of individualized expression. For most of the key concepts of Chinese culture there are important action components and associations in their geometrical and graphic configurations. Perhaps the most central concept in Chinese culture is 'Tao' or 道 . The geometrical component 自 means 'self' and is derived from 目 'eye' or 'to see.' The geometrical component 艹 usually means 'flowers' or 'vegetation,' while the component 辶 means 'to walk' or 'on the way.' The configuration 首 similar to 自 also means 'leader' or 'original.' These composite graphic associations, abetted by the configuration 辶 which is one of the few Chinese word symbol components that has no exact geometrical shape and is neither a continuous curve nor a straight line, is extraordinary and mystic. The 'self,' or ego, ameliorated by the gentleness of 'flowers,' 'central' and 'original,' 'continues' on a 'path' that has no exact direction, nor any clear end. The 'path' 辶 however is an extended geometric form of 夊 which while it contains no meaning as a word symbol on its own contains in it, as a building block component, the 'female' 女 shortened to 夂, which also means 'repetition.' The clear cut geometric shape of 'self' 自 is endowed with a status 自 whose force is ameliorated by the flora expression of 艹 .

The feminized and repetitive way that the self travels is balanced by the mitigated status of self or ego. For the calligrapher, the long, curved line that extends to the southeast allows an expression of force that suggests the unbridled power of libido. Conjoined to the feminized and repetitive part of the figure to the northwest, libido is integrated with superego and the 'female type' ego. In this context the clear 'sight' of the ego, presumably male, is set, and the context of ego-negating and libido-abdicating aspects of superego and the 'female type' in which male ego exists, independent and clear, is further enhanced by the florally ameliorated sense of status or ego recognition which hangs, literally, over his head. The cumulative associations have the effect of emphasizing consciousness, emerging first in an individual form, but ascending to both rational and cosmic levels.

Another word symbol that expresses a key concept in Chinese culture and which allows great freedom of individual expression for the calligrapher is 之 . Formed also of a curved and zigzagged line which has no clear point of termination, this word symbol has a variety of seemingly contradictory meanings and usages. It can mean 'to go to' where it is used as a verb, or 'he,' 'she,' 'it,' 'them' as personal pronoun, or 'of it' as a terminator of a prepositional type phrase. It is best translated at least in most cases where it is used in conjunction with other key concepts of Chinese culture as 'that of which it is the substance of, or belongs to, in the way of its emergence or becoming.' It is also closely associated with the word symbol for 'virtue' 德 which is comprised of the subcomponents, 彳 'action' or 'walking,' 亻 'human being,' 士 'scholar,' 皿 'vessel' or 'utensil' and 心 'heart' or 'mind' placed under 一 'status.' In 之 libido aggression is easily expressed and is ordered only by the single diagonal line that organizes the initial more or less straight line of 'status' 一 , and the bottom, longer line without any clear shape that expresses libido aggression. The word symbol for 'virtue' 德 offers little opportunity for expression of libido aggression except in the curved and heavy line of 'heart' or 'mind' 心. Libido aggression is most clearly expressed not only in the representative but also graphic and geometrical associations that the word symbols or their appropriate subcomponents, signify. Geometrical configurations which suggest ego-recognition and their associations, both representative and otherwise, are usually comprised of straight lines, either vertical or horizontal.

In most of the word symbols expressing key concepts of Chinese culture, the cumulative and ultimate architectonic effect of all the various associations produced by geometrical, graphic, and representational means is the ascendancy and pre-eminence of consciousness as the major resultant solution. Each Chinese word symbol expressing a key cultural concept is like a puzzle, combining in various ways and in fragmented particulars the elements of ego-recognition, libido aggression and superego control. In each puzzle these elements achieve a special balance, and it is an aspect of the calligrapher's art to achieve his own individual interpretation of balance within the ultimate constraints of the puzzle. It is in the balance itself that is achieved the awareness or recognition of the ascendant consciousness as the force which dominates, by moral, practical and logical necessity, all the other forces of the universe. However the basic mode of linguistic expression is based on geometrical and mathematical forms, their associations are ultimately related to psychic forces that are seen on an individual, collective or cosmic level.

The mathematical and geometrical mode in which these associations are expressed allow for a partial and particular differentiation of these levels, but, due to the multidimensional nature of the associations produced, these distinctions are never absolute, even though they can usually be differentiated at least to some extent. Similarly, consciousness on an individual, collective or rational and cosmic levels can be differentiated although not in an absolute way. Balance is achieved in particulars, but its ultimate effect is on a continuum, and true to the realities of human existence, balance on one level ultimately leads to balance on all levels. The underlying theme of the Chinese language system, as a system that is philosophical, psychological and mathematical at the same time as it is linguistic, is that man's cognitive self is inalienable from his instinctual self. Objectification of phenomena modes cannot be separated from subjectification of phenomena modes. In the Chinese language system, the original purposes of the Pharaonic elite has been entirely reversed. In the original 'master code' of the Pharaonic elite, the organization of knowledge for peak 3rd efflorescence society was directed to the purposes of 2nd efflorescence modes. In the Chinese system, as epitomized in its linguistic evolution, the purpose that is implicit in 2nd efflorescence modes has been reorganized and redirected to the ego-negating and libido abdicating philosophies of the late stages of the 3rd efflorescence.

The purposes of empire for which scientific and technical knowledge had been used in the Egyptian civilization, itself a contradictory blend of 2nd, 3rd and 4th efflorescence modes, no longer existed by the time of Fu Hsi when Chinese language was deliberately and self-consciously evolved. By this time some 5-7000 years ago the Egyptian empire itself was near a point of total disintegration, and the surrogate Pharaonic system that had existed in its distant colony on the Chinese mainland had all but collapsed. For the indigenous political and social elite who emerged as a political force in China at that time and who so closely identified themselves with the masses, the libido aggression and ego-recognition modes of the surrogate Pharaonic elite who had ruled the Chinese mainland for so many millenia were loathsome. The intellectual and cultural resources of the 'master code,' which was the legacy of the empire, were used to disseminate the skills of neolithic civilization among the Chinese masses, and the original purposes of the 'master code,' as it had been organized under the direction of the Pharaohs, was redirected and its overall scheme was reinterpreted.

Originally, the wealth of knowledge possessed by the people of the Nile Valley in the peak of 3rd efflorescence society had been organized to the purposes of 2nd and 4th efflorescence modes indigenous to the conquerors of the Nile Valley. It was these purposes which the emerging social and political elite of China eradicated from the 'master code' and a revision of this body of knowledge was undertaken in terms of ego-negating and libido-abdicating religious philosophies compatible with their own indigenous collectives modes and cumulative historical experience. Revision of the 'master code' as an historical event is documented in the legendary account of King Wen which according to Chinese historiography, occurred about 5,000 years ago, about 2,000 years after the time of Fu Hsi.

Like Moses, King Wen was apparently a member of the royal family in the last great pre-historic Chinese dynasty, the Shang, which was ruled by a surrogate Pharaonic elite. Like Moses, King Wen was so distressed by the brutalities and excesses of the Shang that he instigated a rebellion against it as a matter of moral principle. For his pains, he was imprisoned by the Shang rulers, and while in jail undertook a reinterpretation of the Chinese *Book of Changes, I-Ching*, which became the basis of the Chinese ego-negating and libido-abdicating religious philosophies that was the underlying foundation for all subsequent Chinese culture. With the aid of sympathizers within the Shang court, King Wen was released from prison and was able to successfully carry out his rebellion against the Shang, bring it to total defeat, and obliterate the history of its particulars from all subsequent Chinese records. The name bestowed on this legendary individual, King Wen, which means 'King of Culture' was honorific because even after his revolutionary efforts were successful, he refused, due to his own ego-negating and libido-abdicating philosophy, to accept the kingship or even any official position in the new government which he established. Instead his son, again euphemistically known as the Duke of Chou, became king and established the first dynasty considered to be legitimate under the new social and cultural parameters both father and son helped to establish for Chinese civilization.

While the legendary account of Fu Hsi documents the deliberate and self-conscious development of the Chinese language system, the ascendancy of consciousness on all levels over the psychic drives could, through the medium of language, be entrenched only in a symbolic form. This first product of Chinese civilization was developed first of all under the demands for a written language that could be used in the exchange and communication of skills of civilization and their particulars as neolithic civilization developed in China. Its development was the product, not of actively engaged revolutionaries so much as of an emerging, reform-minded, political and social elite among the surrogate Pharaonic elite of China whose learned but not deeply entrenched 2nd and 4th efflorescence modes had begun to seriously dissipate in time and in remoteness from the Egyptian empire and the subsequent turmoil in Near Eastern regions. Under these circumstances, the development of such a writing system was seen as a purely intellectual exercise by the political power structure that still existed in these times, as it posed no direct threat to it. However these early inventors of the Chinese written language system wished to reform the psychic modes of the existing system, their pursuits in these areas were entirely compatible with their offical stations as priests or technicians serving the state. As neolithic civilization developed independently in China, and consequently at a much slower pace than it had in the Near Eastern regions, the political and social effects of this writing system, as a kind of revolutionary act against the surrogate Pharaonic power structure, did not occur immediately. In fact, with the slow dissemination of neolithic civilization in China and the written language system which facilitated its spread, it took several thousands of years for this intellectually revolutionary act to have a social and political aftermath.

Only after neolithic civilization had been widely disseminated and the difficult written language system assimilated as a customary mode of collective and individual behaviour among the masses could any serious social and political change be effected. In harnassing the scientific, technical, and philosophical wealth of peak 3rd efflorescence society, as it existed in the 'master code' the Chinese brought with them from Egypt, to purposes that suited the indigenous conditions of China as they had developed over millenia, the original purposes of this 'master code' was entirely revised. As science and technology had been used in Egyptian civilization for the purposes of a small ruling elite who had substantially different collective modes from the masses they controlled, in China where this was not the case science and technology as such were viewed as irrelevant to the development of civilization. For the purposes of disseminating neolithic civilization among the masses, the sophisticated science and technology which had been used in the service of the particular proclivities of a special collective who in fact no longer existed in China if indeed they ever had, were discarded as irrelevant. The hostility against science and technology which was manifested in Near Eastern regions and India, as they were so closely identified with the brutalities of the Pharaonic elite of Egypt, were however also lacking in China.

As a mere colony of Egyptian civilization on a remote shore it is doubtful whether the more sophisticated and refined aspects of science and technology had ever been delivered intact unto the surrogate Pharaonic elite of China by the paranoid rulers of the Nile Valley. It is doubtful whether the secrets that these authentic Pharaohs of Egypt believed would lead to their physical reincarnation would have been bestowed on mere colonists of the Empire, however important, for economic and strategic reasons, this colony may have been to the Egyptian empire. Thus, for practical as well as philosophical and moral reasons, the scientific and technological, aspects of the 'master code' were regarded as irrelevant in China, in part at least because the more refined and sophisticated aspects of this science and technology were unknown to these distant colonists. The logical and philosophical bases of the science and technology of the 'master code' which was familiar to the surrogate Pharaonic elite in China, as it had been delivered to them intact, perhaps for safe-keeping, could thus be regarded dispassionately in a way that was not possible in Near Eastern regions where the logical and philosophical bases of this science and technology were so closely associated with the science and technology itself and where its application had caused so much misery and suffering for the masses. Thus the indigenous social and political elite of China lacked the moral qualms which might have prevented them from utilizing the logical and philosophical bases of this science and technology for purposes that would benefit the masses and facilitate the dissemination of neolithic civilization.

Unhampered by direct experience with the adverse social and political implications of science and technology, the social and political elite of China undertook to utilize the logical and philosophical basis of science and technology in the methodical development of the Chinese written language system. In a sense the Chinese of 5-7,000 years ago were in the same position as the Greek about 2500-3000 years ago when the logical and philosophical bases of science and technology as the legacy of the Egyptian empire were developed as a form of intellectual exercise and cultural expression. Like the Chinese, the Greeks had had no direct experience with the adverse social and political effects of science and technology under the Egyptian empire and were able to examine their logical and philosophical bases, when they encountered them, with dispassion. The Greeks however had the aim of developing the logical and philosophical bases of science and technology that they encountered to reconstruct science and technology in the service of the collective as a whole, in the service of attaining 4th efflorescence collective modes.

In China, however, this was not the case. For one thing, the Chinese had a far more extensive appreciation of the logical and philosophical bases of science and technology than the Greeks, as they possessed the 'master code' of Egyptian civilization almost intact, while the Greeks, due to their late entry onto the theatre of Egyptian civilization, were entirely unaware of its existence. The people of the Nile Valley in the peak of 3rd efflorescence society had been coerced into organizing the wealth of their knowledge for the purposes of their conquerors. The Greeks who derived themselves from collectives who had passed through 3rd efflorescence society independently of the peoples of the Nile Valley and who had 3rd efflorescence modes well-entrenched into their collective behaviour had never undergone any such experience. In the course of their wanderings, the Greeks had been easily able to pick up the special skills of civilization and their particulars which was the heritage of a long and tumultuous evolutionary process in the Near East whose hazards and miseries they had been happily spared. They were able to cull the end products of this process although they had little apprehension of how this process had been achieved. As a happy accident they were able to select the most favorable aspects of this end process which would enable them to enter full-fledged into 4th efflorescence society.

For the Chinese however, the logical and philosophical bases of science and technology were useful only in so far as they could be used to reverse the process by which they had been achieved in the first place. Although the Chinese may have had little or no direct experience with the adverse social and political effects of the science and philosophy *per se* resulting from the 'master code,' they had had ample direct experience with the adverse social and political effects of the 'master code' in general. Although the more refined and sophisticated aspects of science and technology may have been largely unknown in China, refined and sophisticated forms of social and political control and oppression were well known. The methods by which such social and political control and oppression were exercised, as facets of the social and psychological application of the 'master code,' were of major concern to the indigenous social and political elite of China as they began to identify themselves with the masses under their control. Manipulation of individual psychology and mechanisms of social organization were the major focus for hostility among this newly emerging elite. It was the process by which such manipulation occurred which they attempted to reverse in the act of intellectual revolution which produced the written Chinese language system.

In the 'master code' the logical and philosophical bases for the interpretation of all phenomenological levels was basically uniform. In its most succinct form these logical and philosophical bases for the interpretation of all phenomenological levels existed as a mathematical and geometrical logical system. Among the Chinese, the sophisticated scientific and technical applications of this system, in engineering, electronics and medicine was ill-developed if at all, so that the mathematical and geometrical expression of the logical and philosophical fundamentals which were so closely associated in the Near East with science and technology *per se* and their adverse social and political effects had no pejorative connotations for the Chinese. In this way the mathematical and geometrical expression of these logical and philosophical fundamentals could be used by the Chinese without compunction to reverse systems of psychology and social organization in terms of moral and ethical imperatives. Thus, among the Chinese the logical and philosophical bases of science and technology were used for very different purposes than among the Greeks. The Greeks used the logical and philosophical bases of science and technology that they encountered as bits and pieces to reconstruct systems of science and technology in the service of evolving 4th efflorescence social modes. The Chinese used the logical and philosophical bases of science and technology which they possessed in their entirety, without understanding their application in these areas, to directly symbolize psychological and social mechanisms in the service of evolving a society in the late stages of the 3rd efflorescence.

The version of the Chinese *Book of Changes* or *I-Ching* attributed to Fu Hsi some 5-7,000 years ago and which according to legend appeared to him on the back of a dragon emerging from a river, is that version of the 'master code' which was used by the Chinese to develop their written language system. This version of the Chinese *I-Ching*, like later versions, is composed of symmetrical lines differentiated only by the fact that one kind of line is unbroken (——) and the other is broken in the middle (– –). In another variation of Fu Hsi's *I-Ching* the unbroken line is represented by a small dark circle (•) and the broken line by a small light circle (○). What distinguishes Fu Hsi's version from the later version of King Wen is that the order of these geometrical and mathematical representations is entirely binary. In *I-Ching* these lines are grouped in three's, and in sixes (or two groups of threes), for a total of 8 orderings of groups in threes and 64 orderings of groups in sixes. In Fu-Hsi's version the total set of 64 orderings of groups in sixes is organized by binary logic, and Leibnitz's 'discovery' of the binary system and also calculus in Germany in the late 17th century is attributed to the inspiration he derived from the binary logic of Fu Hsi's version of the *I-Ching* some 5-7,000 years ago.

Thus the order basically proceeds in the simple way

etc, and in modern Boolian algebra can be represented in the more familiar fashion of 01, 00, 111, 011, 001, 000, etc. This purely mathematical, geometrical and arithmetical version of the *I-Ching* or 'master code' represents the singular use to which it was put by the inventors of the Chinese language system. Expressed in binary logic, these linear forms were modeled in many geometrical configurations, perpendicular, horizontal, diagonal, and circular. The logical and philosophical bases of this mathematical, geometrical and arithmetical system was used to develop the building blocks of the Chinese word symbols through the mental associations they created.

The means of evolving these associations were geometrical and mathematical and depended on a complete understanding of the logical and philosophical bases of geometrical and mathematical systems. But the particular associations that these means were used to create were invariably of a psychological and sociological ilk. The psychic drives of man in an individual, collective and cosmic context were represented. The means by which they were represented, geometrical and mathematical, produced abstract representations of these drives so that it was difficult if not impossible to clearly separate their particular focus from one level, individual, collective, or cosmic, to another. As in Greek culture, the Chinese written language system created a culture where man rather than God was the center of the universe. However unlike in Greek culture where 'man was the center of all things' because of his ability to objectify phenomena, in Chinese culture man was the center of the cosmos because he was able to attain awareness of his ability to subjectify phenomena. Or to put it another way, in Greek culture 'man was the measure of all things' because he could objectify the subjective, while in Chinese culture 'man was the measure of all things' because he could subjectify the objective.

While the invention of the Chinese written language by the indigenous social and political elite of China was a revolutionary act of an intellectual, rather than political nature, and as neolithic culture became disseminated in China with the aid of this language, it was not sufficient to the aims of this new elite, as typified by King Wen, who desired an overall change in the total political and social structure of Chinese society. Although an ego-negating and libido-abdicating religious philosophy had been built into the written Chinese language system, the continuing existence of a surrogate Pharaonic elite as the ruling political and social structure of China represented 2nd and 4th efflorescence modes which openly acknowledged ego-recognition and the expression of libido aggression as the model of social and political authority. Practices of religion and magic which involved a symbolic if not a real belief in the physical reincarnation of the ruling surrogate Pharaohs permitted the continuation of cruel and brutal practices against the hapless masses.

Political and social revolution was seen as a necessity whereby the religious philosophies of ego-negation and libido-abdication could be instituted, but the written language system itself was an insufficient vehicle for such institutional changes. In order to develop a new theory of state and government which would enable political and social revolution to become a reality, the emergent indigenous social and political elite of China found it necessary once again to review the 'master code,' and to revise it on a more intensive scale in order to devise a new philosophy of state and government. Thus, the version of the Chinese *Book of Changes* or *I-Ching* attributed to King Wen about 3-5,000 years ago, who was responsible for the overthrow of the Pharaonic system of state and government, is no doubt closer to the original 'master code' than that of Fu Hsi's some 5-7,000 years ago.

For the purposes of inventing the Chinese written language, only certain aspects of the 'master code' were necessary, those that involved the logical and philosophical bases of the scientific and technological aspects of the 'master code.' For these purposes the presentation of the *I-Ching* as a binary system of mathematics, geometry and arithmetic are sufficient. In devising a philosophy of state and government, however, for purposes of dismantling the Pharaonic social and political system, it was necessary to understand in its entirety the basic psychological, social and religious mechanisms that made this system operative. It is clear from the dynamics of the Chinese written language that its inventors understood clearly the psychological mechanics which impel man to action, on an individual, collective and cosmic level. From the point of view of the Chinese written language system, as in modern theories of psychology, individual psychic drives are in all cases prior to collective drives, which in turn are prior to cosmic drives. Collective psychic forces were seen as arising from individual psychic forces as an extension of them, and cosmic psychic forces in their turn were an extension of the collective psychic forces, where the center of these drives was always within the individual himself and not displaced in some being or state outside of the individual.

Yet, in terms of the actualities of collective and religious action, the individual tended to behave as if this were not the case, as if the forces that impelled him in these spheres of action were diffused in the collective itself or in the cosmos. For all the efforts made in the invention of the Chinese written language to enable the individual to understand that collective consciousness or unconsciousness as such, or cosmic consciousness or unconsciousness as such, were merely cognitive and emotional extensions of his own individual drives, it was apparent from the actualities of collective behaviour that the individuals of the collective did not behave as if this were true. The Chinese written system was an insufficient method by which the individuals in the collective could contain and control their social and political behaviour. However well-entrenched the Chinese written language system had become by 5-3,000 years ago, members of the collective continued to demonstrate an irrational fear and awe of the surrogate Pharaonic elite and the religious concepts which they symbolized, and against all precepts of rationality and balance, behaved like automatons in the face of certain key signs and signals from this elite, leading to the same brutal social and political excesses that they had engaged in prior to the invention of the Chinese written language system.

The emerging indigenous social and political elite felt compelled to re-examine the 'master code' more extensively in order to discover how the social and political behaviour of the collective was structured on a mass scale, in order to devise a theory of state that would allow it to be restructured. Thus, the King Wen version of the Chinese *Book of Changes*, or *I-Ching*, provided the direct inspiration for the work of Confucius, who attained in Chinese civilization the philosophical status of Socrates in Greek civilization, high priest as Moses in Hebraic civilization, and Buddha as the model of exemplary moral living in Indian civilization, and Confucius' commentaries of King Wen's version of *I-Ching* are regarded in China, along with those of King Wen himself, as the final word on the 'master code.'

Chapter XVI

CONFUCIANISM VERSUS TAOISM: ANCIENT CHINA STRUGGLES TO CREATE A PHILOSOPHY OF STATE

The religious philosophies of China, as exemplified by Confucianism and Taoism both about 2500 years old, recognize the source of collective action and cosmic consciousness in individual psychic drives. The difference between the recognition in Chinese civilization of the mechanisms of social organization and collective action, as well as cosmic consciousness, as 'man-centered' and the same recognition in Greek civilization is a difference in the degree of precision. Although the Greeks too recognized that man was the center of collective action and cosmic consciousness, by comparison with the Chinese, the Greek concept of man himself was rather vague. This ambiguity about the exact nature of man in Greek civilization permitted an easy and rather facile acceptance of collective action and consciousness without a great deal of introspection that was not possible in Chinese civilization. Thus, in Greek civilization modes of both 2nd and 4th efflorescence society were accepted pragmatically: they were taken for granted because in Greek civilization they happened to work, and work without inhibiting developed modes of 3rd efflorescence society, or entrenchment of superego.

Only in late Roman society, which had adopted the Greek model of civilization, where 2nd and 4th efflorescence modes were seen to seriously inhibit 3rd efflorescence modes, was Christianity developed and adopted, as a way of maintaining working 2nd and 4th efflorescence modes with reinforced superego entrenchment as personified in the life and teachings of Jesus Christ. Indian civilization too recognized the center of collective action and cosmic consciousness in human psychic drives, but due to historical experience never perceived any practical utility for collective action as an aspect of 2nd and 4th efflorescence modes. Thus, Indian civilization too, for different reasons, lacked a sense of precision about the location of collective action and cosmic consciousness as aspects of individual psychic drives, as in Indian civilization both collective action, and individual psychic drives in tandem with collective action, were regarded as irrelevant.

Due to the extraordinary influence of the 'master code' on both the Chinese written language and the later developed systems of religious philosophies which doubled as systems of state and government, in Chinese civilization the individual tended to be regarded from a technological point of view, rather than a philosophical point of view as in Greek civilization, or a mystical point of view as in Indian civilization.

In Chinese civilization the view of the individual as a 'machine' whose components could be precisely analyzed went beyond the latter day theories of both Freud, Jung and Marx. As a 'machine' that could be precisely developed in terms of his component parts of psychic drives, man was a 'machine' that could not only be precisely analyzed but also precisely controlled. The logical conundrums of such a view are a matter of historical fact, and at every turn Chinese civilization as the epitome of the model of the late stages of 3rd efflorescence society stands in stark contrast to the model of 4th efflorescence society derived from Greek civilization in the kinds of solutions it offers to the human predicament. For one thing, in theory at least, Chinese civilization offers the potentiality of a completely classless society, devoid of institutionalized slavery in any form, which 4th efflorescence society in both theory and practice has yet to come to terms with. In actuality, however, this theory has always remained a potentiality rather than a reality in Chinese civilization, and Chinese civilization on the whole has been less successful in creating an egalitarian society than at least some known and proven instances of 4th efflorescence modes based on the model of Greek civilization. The fact that Chinese civilization remains a model that is perfectly accessible to rational analysis and yet does not produce what it promises reveals the hiatus between man's power of rational thought and actual reality.

In Chinese thought the individual is comprised of the four basic drives, ego ('heaven'), libido ('yang'), superego ('earth'), and consciousness ('yin'). These basic individual psychic drives extend into collective and cosmic drives. Theoretically, ego-negation and libido-abdication will result in the lessening of the power of those drives which prove destructive on the individual, collective, and cosmic levels. Recognition of ego and libido is achieved through consciousness ('yin') and attenuated by superego ('earth') and this recognition or awareness will result in a total balance or harmony of all drives or 'Tao,' for the general benefit of the individual, the collective, and the cosmos. When this harmony is achieved—the *chung yung* or 'middle way' of Confucius—the ascendant drive will be consciousness ('yin'), albeit not on an individual or collective level, but on the higher cosmic level. All the individual drives, including individual consciousness, become subsumed in cosmic consciousness which is balanced by cosmic libido, and the ultimate 'Tao' or 'Tai Chi' is achieved. Once *chung yung*, 'tao' and 'Tai Chi' are achieved by every individual, under his or her own direct individual control, the problem of state and government will be solved, since in fact at this point there will no longer be a need for any formal or institutionalized state or government. Collective action will not result in any excesses or brutalities since each individual operates under strict moral imperatives which force him to regard the welfare of every other individual as equal to his own.

These are the basic teachings of religious philosophies doubling as philosophies of state and government which comprise the logos of Confucianism and Taoism, and from a strictly logical point of view it is impossible to fault them. Although they have in fact never actually worked, never produced what they promise, does not logically prove that under the right circumstances at some point they might in fact work and produce what they promise. The experiential reality of their failure does suggest that somewhere in this precise logical system there might be a fallacy, however difficult it might be to discern it. Of course to call into question the logical efficacy of such a system is also to call into question the logical efficacy of such modern thinkers as Freud, Jung, Marx whose technological view of man also promises solutions for individual and social excesses and institutionalized forms of slavery in all of its various modes. This is not to argue that a logical analysis of man and society is counterproductive but rather to suggest that when the biological reality of man's evolution is taken into account, purely logical analyses of man's condition present certain unfortunate caveats.

When we look at the logical analysis of man and his condition given by Chinese religious philosophies, which in their precision makes the works of modern thinkers such as Freud, Jung and Marx and others seem rather insipid by comparison, we may realize that a logical *summa* of all that we believe to be 'modern' thought has already been in existence for many thousands of years. From a mundane point of view, we may see that our logical knowledge, if not our practical progress, has advanced little over these thousands of years. This fact alone is sufficient to explain why societies in the so-called 'underdeveloped' world seem so reluctant to acknowledge the superiority of the so-called 'developed' world. They remain abashed at the hubris of man in 4th efflorescence society who pretends to have solutions to problems he cannot even properly formulate. In societies in the late stages of the 3rd efflorescence, in the so-called 'underdeveloped' world as epitomized by Chinese civilization, there is a much clearer understanding than there is in the 'developed' world that collective action is centered in individual psychic drives.

Until a method is found to govern individual psychic drives on a mass scale, collective modes, which are so valued in 4th efflorescence society, remain only an extravagant displacement of individual psychic drives, out of the ken of individual consciousness and control and ultimately destructive to all individuals and the collective as a whole. From this point of view, the technological, economic and scientific progress of the so-called 'developed' world is merely a chimera, tempting mankind into more and more extravagant displacements of individual ego and libido into an unconscious and uncontrollable collective ego and libido that becomes ever more threatening to human survival. Again, while this argument seems logically flawless, it goes against the deepest human and biological drives of all—evolutionary progress, and it is defied by the visible and practical advantages that 4th efflorescence society, for all of its logical flaws, has brought to mankind as a whole in terms of the general welfare and betterment of human life in most, if not all, instances.

Thus, if there are flaws in the logic of the religious philosophies of societies in the late stages of the 3rd efflorescence as epitomized by Chinese civilization, it must be in the basis of logical and philosophical thought itself, the intricacies of which, for better or for worse, were only skirted by the Greeks and never understood with the precision or the depth of the Chinese. The first question is whether or not the basis of logical and philosophical thought of the Chinese is limited by the source of its origin in the ancient Egyptian 'master code.' The second question is that even if it is limited by the source of its origin, whether or not the basis of all logical and philosophical thought will have the same ultimate limited end. That is, if as the basis of logical and philosophical thought in 4th efflorescence society expands in breadth and depth, in the process of evolution, will it attain the same logos of knowledge contained in the Egyptian 'master code' from which the Chinese derived their religious philosophies? In other words, is logical thought a closed system, and will it always attain the same end, the logos of knowledge contained in the Egyptian 'master code' of which the Chinese *Book of Changes* is the only extant version? If we admit that logical thought is a closed system, we may be prevented from attacking our serious social and political problems with the tools of rational analysis, and thereby come to the same unhappy end predicted for 4th efflorescence society in Chinese religious philosophies. If we deny that logical thought is a closed system, we may with undue alacrity accept the solutions offered by the religious philosophies of Chinese and other 3rd world civilizations and bring our process of social, cultural, and human evolution to an untimely end of stagnation and decay.

In fact, the solution is one of compromise. In all great civilizations of both the late stages of the 3rd efflorescence and the 4th, logical systems of thought that work in practice can never be completely closed systems. Even in the religious philosophies of China, the concept of 'Tao' is an open concept which in the end defies all laws of logical analysis. In Near Eastern civilizations the open concept is that of 'God,' and in Christianity it is that of 'God' and Jesus Christ. Even in the social systems of Marx and Hegel the open concept of 'history' as a force exists, and in the theories of Freud and Jung all the human psychic drives remain open concepts as they expand into the collective and cosmic arenas. In Darwin 'evolution' is the open concept of the 'unknown.' The apparent closure of logical thought in religious philosophies of the late stages of the 3rd efflorescence, particularly as epitomized in Chinese religious philosophies, results from too close an identification of the levels of the individual, collective and cosmic forces, as they emanate from the individual psychic drives. Although seemingly logically precise and undoubtedly mathematically and logically aesthetic, this too close identification of these different phenomenological levels, individual, collective and cosmic, are unrealistic.

As the basis of logical and philosophical thought, this too close identification of these different phenomenological levels derives from the original purpose for which the 'master code' of the Egyptians was devised, and which in one way or another has profoundly influenced all societies of both the late stages of the 3rd efflorescence and the 4th which have come down into modern times. The original purpose of this 'master code' devolved on the powerful 2nd efflorescence modes of the Pharaonic elite whose egos were so extravagant that they demanded that the entire cosmos be anthromorphicized in their own image. Collective and cosmic modes as apprehended by peak 3rd efflorescence society were personalized in the humanoid image of psychic drives as personified by the Pharaonic elite, in particular the single God-King, the Pharaoh.

Such personalization of collective and cosmic modes allowed the Pharaonic elite to envision the universe under their direct control, and deluded them into believing that their own egos could be made eternal through the possibility of physical reincarnation. Logic, as an aspect of science, evolved as a 2nd efflorescence mode, and logical completeness is able to exist only under the limiting conditions of this evolutionary mode. This is not to say that logical completeness can be discarded simply because it is a limiting condition of the 2nd efflorescence mode, anymore than ego can be discarded for the same reason. Like ego, logical completeness is a facet of man's evolutionary development. It cannot be discarded, but must be recognized for what it is, a limiting condition of man's evolutionary development, with proper adaptive advantages when treated with discretion and recognized in the fullness of its evolutionary potential for change.

Reasoning by a series of analogous steps which anthromorphicize the universe is characteristic of the Chinese religious philosophies and is revealed in the famous line by Confucius', 'He who first makes his heart sincere, can control his own self (body and mind); he who controls his own self, can control his family; he can control his family, can control the nation; he who can control the nation can control the universe.' In this mode of reasoning all the wealth of knowledge derived from subjectification of phenomena modes in the entrenchment of superego and that were then objectified as special skills of civilization and their particulars are re-subjectified. From a psychological point of view the utility of such a mental procedure is that the individual is able with clear precision to identify all the projections of personality that occur as objectifications and use these objectifications in the service of internal psychic integration and coherence. The end ideal results in a perfectly balanced individual who has phenomenal control over and integration of his own involuntary and voluntary nervous systems. Although in the texts of Taoism, the concept of 'Tao' itself remains an open ended concept, as it cannot be seen, or named and has no image of any kind, 'Tao' can only be achieved after this process of personalizing the universe and locating it within the individual psyche is first mastered.

Thus the texts of Taoism explicates the way in which the 'unnamed' 'Tao' can be attained only through first understanding this process of 'naming.' In the *Tao Te Ching* all those things which can be 'named' are thrown into the same phenomenological class, to be distinguished from that which cannot be 'named', i.e., 'Tao':

> The name that cannot be named is not the eternal name.
> The nameless is the beginning of heaven and earth.
> The named is the mother of ten thousand things
> Ever desireless, one can see the mystery.
> Ever desiring, one can see the manifestations.
> These two spring from the same source but differ in name;
> this appears as darkness . . .* (Chapter 1)

The ease with which all phenomenological levels can be fused into the same general class in the service of subjectifying the objective and thereby attaining balance or 'Tao' is revealed clearly in the following lines:

> In the pursuit of learning, every day something is acquired.
> In the pursuit of Tao, every day something is dropped. (Chapter 48)

All quotations from the *Tao Te Ching* are from the translation by Gia-Fu Fung and Jane English, Vintage Books, 1972, New York.

However the price paid for this perfect psychological model of individual balance is passivity or neglect in terms of both collective and even cosmic modes:

> Less and less is done
> Until non-action is achieved.
> When nothing is done, nothing is left undone.
> The world is ruled by letting things take their course.
> It cannot be ruled by interfering. (Chapter 48)

Of course there is a clear awareness in the Chinese religious philosophies that phenomenological levels of awareness are in reality different, as is revealed by:

> Therefore look at the body as body;
> Look at the family as family;
> Look at the village as village;
> Look at the nation as nation;
> Look at the universe as universe.
> How do I know the universe is like this?
> By looking! (Chapter 54)

Levels of phenomena are however fused because it serves the ends of individual balance, or 'Tao,' which can then be re-objectified to all these various levels of phenomena:

> Cultivate virtue in yourself,
> And Virtue will be real.
> Cultivate it in the family,
> And Virtue will abound.
> Cultivate it in the village,
> And Virtue will grow.
> Cultivate it in the nation,
> And Virtue will be abundant.
> Cultivate it in the universe,
> And Virtue will be everywhere. (Chapter 54)

By comparison with Confucianism, Taoism seems almost flippant, as it chooses to see collective modes entirely from the point of view of a single 'sage,' whose complete attainment of 'Tao' will somehow permit it to be re-objectified in all of the collective and cosmic modes. It is therefore no accident that Confucianism rather than Taoism became the offical philosophy of state in China for 2500 years, since, unlike Taoism, Confucianism was loathe to place the destiny of an entire nation in the hands of a single Taoist sage, however accomplished and balanced he might be. The tenets of Confucianism rested on the fact that moral virtue, which might eventually but not always lead to 'Tao,' could and should be universally inculcated as a matter of education in all members of the collective. Thus, Confucianists find particularly flippant and irresponsible the following passage from the Taoist texts which mark the single most critical idological split between Taoism and Confucianism:

> Therefore when Tao is lost, there is goodness.
> When goodness is lost, there is kindness.
> When kindness is lost, there is justice.
> When justice is lost, there is ritual.
> Now ritual is the husk of faith and loyalty, the beginning of confusion. (Chapter 38)

These basic precepts of 'goodness,' 'kindness,' 'justice,' and 'ritual,' are among the most central tenets of faith taught by Confucianism, tenets that unlike 'Tao' could be universally propogated and clearly understood and applied in concrete situations. As a matter of the ideological split between Taoism and Confucianism the next line in this passage refers to the Chinese *Book of Changes*, *I-Ching*, to whose study Confucius devoted much of his life and from whose teachings he admittedly derived much of his thought:

> Knowledge of the future is only a flowery trapping of Tao.
> It is the beginning of folly. (Chapter 38)

The *I-Ching* was thought of by Confucius' time and by the time of the author of the Taoist texts as a book of divination or prediction, and Confucius' particular explications of the *I-Ching* did little to dispel that notion.

The fact is of course that while both King Wen and Confucius often used the *I-Ching* or the 'master code' to develop their sophisticated religious philosophies and philosophies of state, they had little stomach for what the *I-Ching* actually represented, and the interpretations they presented of the *I-Ching*, as they have been passed down in the written literature, did more to obscure the actual purpose of the *I-Ching* than to enlighten them. Through their efforts, the actual purposes of the *I-Ching* or 'master code' were deliberately obscured if not suppressed. That there existed a serious ideological rift on this matter of obfuscation is seen in the next line of the Taoist texts:

> Therefore the truly great man dwells on what is real
> and not what is on the surface,
> On the fruit and not the flower.
> Therefore accept the one and reject the other. (Chapter 38)

The authors of Taoism obviously felt there was a substantial virtue in the *I-Ching* or 'master code' which did not deserve obfuscation, but enlightenment. That the general consensus among the Chinese political and social elite was that they were wrong in this is attested to by the fate of Taoism as a governing philosophy in Chinese civilization. In most periods of Chinese dynastic rule Taoism was considered a subversive philosophy, and followers of the Tao were confined to monastic life outside of the mainstream of all official social and political institutions. Taoism became a religious philosophy that was institutionalized in secretive and cultist modes, and the method of transmission of Taoism was largely oral rather than written.

As a religious philosophy that remained largely underground but which would remain popular among common folk in China even into modern times, many of the vestiges of the 'master code' were kept and deified, since among the illiterate masses of peasantry who were unable to afford the lengthy and difficult training which Confucianism proper demanded, these vestiges were a holdover from early neolithic times when skills of civilization had been disseminated under the auspices of the surrogate Pharaonic elite some 5-7,000 years ago. Even today in Taoist funerals in Taiwan and Hong Kong, and even mainland China, funerary rituals, compatible with the financial resources of the masses, are a replica of ancient Egyptian burial rites where the dead personage and his retinue and household goods are represented by models made of paper or clay. The paper or clay models are burned, signifying the passage of the dead with his retinue and household goods on their way to the underworld, through which they hope to pass unscathed on their journey from death to life again, where they may be reincarnated. In fact, much of the knowledge that still exists today, about the actual structure and purpose of the ancient 'master code,' of which the Chinese *I-Ching* is an extant version, comes through the oral tradition of Taoism as it has survived in formalized religious modes and associated cults and societies in China. For the pragmatic Confucianists who hoped to use the *I-Ching* for the purposes of building an independent civilization in China with original social and political institutions, the substance of *I-Ching* was seen in a particular light. Although it was considered worthy of veneration due to the ancestral roots of the new Chinese civilization in *I-Ching*, for a variety of reasons serious investigation and clarification of it was discouraged.

The familial pattern as a model of state which was taken from *I-Ching* by logical necessity demanded that *I-Ching* be venerated, since the modes of analogous reasoning characteristic of Chinese culture would classify ancestral culture in the same general level as familial ancestry, or ancestor worship, a mode of the religious philosophy of evolved Confucianism. However, just as all ancestors were considered worthy of worship for their contribution to familial propogation and the propogation of the race, and the sins of these ancestors genteelly forgotten or suppressed, so the *I-Ching* was considered worthy of veneration for its contribution to Chinese culture as a matter of the propogation of that culture, while the particulars of its substance with all of their negative associations were conveniently obscured. For one thing, resuscitation of the substance of *I-Ching* would have called to light the unhappy historical circumstances in which Chinese civilization had its original roots.

The extraordinary originality and creativity of the Chinese written language system and its aftermath in Chinese religious philosophies in fact owed but little to the dim past of the Chinese collective whose roots lay in another place and time long since forgotten even in its earliest stages of unique evolution on the East Asian mainland. Over a period of thousands of years the surrogate Pharaonic structure which the original settlers of China had brought with them from the Nile Valley underwent a transformation that left the emerging social and political elite of China at odds with this inherited structure at every turn and at every point. Only grudgingly did this newly emerging indigenous social and political elite accept the logical and philosophical benefits of the 'master code' to develop the ego-negating and libido-abdicating concepts that went into the development of the Chinese written language system, itself a relatively gentle act of revolution against the main thrust of the 'master code' and all that it represented in the social, political and religious spheres.

The remaining structures of surrogate Pharaonic modes with their social, political and religious excesses and their lack of empathy with the masses continued to be abhorrent to this newly evolving elite, and resulted in a genuine social and political revolution against the state in the time of King Wen. In fact, the predicament of this genuinely lively and inventive people on the Chinese mainland, so conveniently isolated from the turmoil in the Near East, was no less than the predicament of other peoples heir to the legacy of the Egyptian empire in the Near East, India and the West. But isolated as the Chinese were from the other civilizations with whom they shared a common thread in the origins of their civilization under the aegis of the Egyptian empire, the Chinese, more easily than these other collectives, were able to maintain the illusion that they had been the original creators of a civilization entirely peculiar to themselves. Cut off as they were from a direct historical consciousness of the origins of the 'master code' which they had in their possession, the substance of that 'master code' encouraged, rather than detracted from, the idea that they were the unique creators of civilization peculiar to themselves alone. Oriented and directed as the 'master code' was to 2nd efflorescence modes, the revision of that code in terms of ego-negating and libido-abdicating concepts and philosophies propogated the sense of freshness and originality with which mankind left behind 2nd efflorescence modes in the course of evolutionary progress and emerged in a later, more advanced stage of evolutionary development.

The unique historical circumstances under which the genesis of Chinese civilization occurred bolstered the illusion that historical and evolutionary development were a single process. In the light of the image that the Chinese had of themselves, due in no small part to the logical and philosophical benefits they culled from the 'master code,' it seemed as if they had evolved, by virtue of their own self-conscious intellect and will, from an amorphous primitive state of 2nd and 3rd efflorescence modes directly to an advanced state of civilization whose boundaries were self-limiting. In the late stages of 3rd efflorescence society which they had achieved, they could regard 4th efflorescence modes in the same sceptical way in which they regarded 2nd efflorescence modes, as limitations on man's individual, collective and cosmic modes of behaviour that they had already mastered.

Once the original purpose of the 'master code' in its historical and evolutionary context is understood, it is possible, with the aid of interpretative knowledge passed down through Taoist schools of thought in China, to use the King Wen version of the Chinese *I-Ching* to reconstruct the system of thought it contained. Interpretations of *I-Ching* by Confucius and others who followed Confucius in his particular line of investigation, although constrained by the moral imperatives inherent in the religious philosophy peculiar to Confucianism, derived much of their special knowledge of *I-Ching* from the Taoist schools, as those interpreters demanded a substantive rather than formal approach to *I-Ching*. Thus, the written versions of *I-Ching* interpretations by both King Wen and Confucius and those that followed their particular method of written interpretation, can nonetheless be regarded as more in the Taoist than in the Confucianist lines of thought proper. From these interpretations it is possible to place the particular emphases of Confucianism in the context of the *I-Ching* or the 'master code.' Interpretation of *I-Ching* thus falls, as it were, on neutral ground in the ideological battle between Confucianism and Taoism in the history of Chinese civilization, since both the written interpretations of *I-Ching* and the oral interpretations of Taoists are blended in this task. Ironically, such a reconstruction of the 'master code' in a way that would be comprehensible to 'developed' societies would have been impossible even 100 years ago, as modern science, psychology and philosophy in the so-called 'developed' societies did not possess prior to that time the sophisticated concepts necessary to an understanding of it.

Chapter XVII

TABULATION OF CIVILIZATION SKILLS IN THE *I-CHING*

The basic components or building blocks of *I-Ching* are broken (– –) and unbroken (——) lines, which are regarded in Chinese written literature as + ('plus') for unbroken lines and - ('minus') for broken lines. This is equivalent to the 0,1 notation for modern computer science, where '0' signifies the computer terminal switch is 'off' and '1' signifies that it is 'on.' Whereas in modern mathematical theory and computer science we are accustomed to envisioning this '0' and '1' in units of two or binary units or 'bytes,' in the *I-Ching* they are grouped in threes. For 0 and 1, or +, - in groups of threes there are 8 possible combinations: +++, ---, +-+, -+-, --+, +--, ++-, -++ (or 111, 000, 101, 010, 001, 100, 110, 011). In the classical *I-Ching* this notation is written:

Although there is a complete literature with symbolic meanings for all eight of these groupings with religious, psychological, sociological, scientific, etc., associations, the first four, which are the most important, have the meaning of ☰ heaven; ☷ earth; ☵ yin (moon); ☲ yang (sun). These notations in threes in the complete *I-Ching* are combined in 2 groups of 3's as in

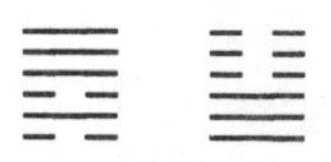

etc., for a total of 8 x 8 = 64 combinations or 64 hexagrams. For these 64 hexagrams there is a mathematical probability of 64! arrangements of 1.2688692×10^{89} possible arrangements, a number that approaches the numerical limit of the universe. The fact that only two arrangements of these 64 hexagrams have been considered legitimate, one being the binary arrangement of Fu Hsi, and one that of King Wen, is itself a matter of extraordinary mathematical interest, and suggests the ultimate scientific uses of this ancient system as a computer and mathematical matrix representing a master codification of an enormous amount of information.

Rather than trying to understand it on the phenomenological level of science or technology *per se* which purposes it no doubt also possessed, although these purposes were obscure even to the ancient Chinese interpreters of the system, we may use the modern psychological terminology in whose terms this system may be comprehended today and in whose general sense it was understood by its ancient Chinese interpreters. Using the psychological concepts and terminology of ䷀ heaven for ego, ䷁ earth for superego, ䷾ yin for consciousness, and ䷿ yang for libido, we may see how the system of 64 hexagrams composed of these basic building blocks develops an intrinsic logic of its own. This system of logic produces an ontological and epistomological system or matrix of knowledge whose components in the proper order, concomitant with King Wen's scheme, possess a hierarchical order of all the skills of civilization and their particulars.

In the eight rows and eight lines of 64 hexagrams comprising the King Wen version of the *I-Ching*, the system is read from right to left, and from north to south as in the Chinese written language system. Starting with the row on the extreme right (see p. 273), alternate odd rows (1, 3, 5, 7) are considered dominant while the even rows (2, 4, 6, 8) are considered latent. However, the King Wen version of the *I-Ching* is also considered to be divided laterally in two halves, a top grouping of 32 hexagrams as opposed to a bottom grouping of 32 hexagrams.

In the bottom grouping of rows (1, 2, 3, 4, 5, 6, 7, 8) and lines 5, 6, 7, 8 there are no dominant or latent rows, both odd and even rows in this bottom half of the system being considered equally active and/or inactive. On the whole, the top 32 hexagrams, read from north to south, are considered to represent categories of knowledge or skills of civilization that are more general, while the bottom 32 read from north to south as well are considered to represent categories of knowledge or skills of civilization that are more particular. Thus the kind of knowledge in the top half is more substantive, essential, and has a greater tone of subjectification of phenomena modes, while in the bottom half knowledge is more abstract, specific, and has a greater tone of objectification of phenomena modes.

From the point of view of the psychic concepts and terminology we are using to interpret the *I-Ching*, the top half represents psychic forces in their individual modes, while the bottom half represents psychic forces in their collective modes. Thus the odd rows in the top half represent the individual consciousness, and the even rows represent individual unconsciousness. In the bottom half the rows represent collective conscious, which is divided by the collective unconscious in odd rows and the collective conscious in the even rows:

uncons.	cons.	uncons.	cons.	uncons.	cons.	uncons.	cons.	
8	7	6	5	4	3	2	1	indivi-
							2	dual
							3	psyche
							4	
cons.	uncons.	cons.	uncons.	cons.	uncons.	cons.	uncons.	
							5	collec-
							6	tive
							7	psyche
							8	

While the difference between conscious mind and the unconscious mind is extremely significant in the realm of the individual, this is not the case in the realm of the collective psyche, where both collective unconscious and collective unconscious may be either equally accessible or inaccessible. This explains the dominant/latent dichotomy in the top half of the *I-Ching* and the lack of such a dichotomy in the bottom half. It is easy to see here how under conditions that existed in ancient civilizations the top half of the 'master code' would gain associations with life and the bottom half with death, or more precisely, how the upper half would take on associations with the 'upper world' or 'world of life' while the bottom half would take on associations with the 'lower world,' 'under world,' or 'world of death.' In general these two halves may be considered the division between the eros and thanos of modern psychological theory and terminology. Given

KING WEN ARRANGEMENT OF THE 64 *I CHING* HEXAGRAMS AS LISTED NUMERICALLY IN THIS VOLUME OF THE TEXT.

8	7	6	5	4	3	2	1
PI	SHIH	SUNG	HSU	MÊNG	CHUN	K'UN	CH'IEN
16	15	14	13	12	11	10	9
YU	CH'IEN	TA YU	T'UNG JEN	P'I	T'AI	LU	HSIAO CH'U
24	23	22	21	20	19	18	17
FU	PO	PI	SHIH HO	KUAN	LIN	KU	SUI
32	31	30	29	28	27	26	25
HENG	HSIEN	LI	K'AN	TA KUO	I	TA CH'U	WU WANG
40	39	38	37	36	35	34	33
HSIEH	CHIEN	K'UEI	CHIA JEN	MING I	CHIN	TA CHUANG	TUN
48	47	46	45	44	43	42	41
CHING	K'UN	SHENG	TS'UI	KOU	KUAI	I	SUN
56	55	54	53	52	51	50	49
LU	FENG	KUEI MEI	CHIEN	KEN	CHEN	TING	KO
64	63	62	61	60	59	58	57
WEI CHI	CHI CHI	HSIAO KUO	CHUNG FU	CHIEH	HUAN	TUI	SUN

King Wen arrangement of the 64 I Ching hexagrams

the fact that this system was conceived in peak 3rd efflorescence society where religious awareness was rich and intense, the lower half, 'underworld,' does not necessarily lend itself to the facile designation of 'hell' since 'underworld' or 'world of death' might be either 'heaven' or 'hell,' depending on the particular perspective. Systemically, however, the dichotomy involves the difference between 'the world of life' and 'the world of death' which is very much in accord with the kind of supernal divisions common to religions of ancient civilizations in Egypt, Near East, India, Greece, and China, rather than to the religions that developed much later in some of these areas and which had clear notions of 'hell' as opposed to 'heaven' such as Christianity and Islam.

According to methods of interpretation passed down in the oral traditions of the Taoist school and schools of Confucianist thought that specialize in *I-Ching* interpretation, read from right to left and from north to south beginning with row 1, line 1, each hexagram is not only a category of knowledge but also a source of error. Each subsequent hexagram (as row 2, line 1 is subsequent to row 1, line 1 and so forth) presents a new category of knowledge and provides a means of correcting the error in the previous category. This new hexagram also however is a source of error of a different kind which is corrected by the subsequent hexagram (as in row 3, line 1). In this way a new sequence of knowledge, correction, and error is set up by each new hexagram, and the relationships between hexagrams proceed or evolve in proper dialectical fashion. However, the phasic alternation between the conscious and the unconscious, or the individual and the collective, or eros, and thanos, life and death, is maintained in the alternation of odd and even rows from right to left. Thus in row 1, line 1, the category of knowledge involves the individual psyche, eros, life, and its errors are corrected in row 2, line 1 by a category of knowledge that involves the collective psyche, thanos, death. Thus the errors that are corrected by the category of knowledge row 3, line 1 that involves individual psyche are the summation of errors in both row 2, line 1, (collective psyche) and row 1, line 1 (individual psyche) and so forth.

From the point of view of logic alone we can see how 'triagrams' (+, - in groups of 3's) and hexagrams (2 groups of 3's) become necessary in such a system. If the first category, line 1, row 1, represents the individual, eros, life, we may give it a single +. Its correction in line 1, row 2, may be represented by a single - ('minus'). The correction of these two categories which are directed to correct the error of the previous two categories with emphasis on the most immediate previous one, i.e., line 1, row 2, gives the next category, line 1, row 3, a designation of - +. The correction of this error in line 1, row 4 may be represented by + -. For correction of line 1, rows 3 and 4 in line 1, row 5, the representation necessarily becomes + - +.

However, the correction in line 1, row 6 cannot be represented by - + - for this would confuse it with line 1, row 3; if it were represented by - + + it would be confused with line 1, row 3 again. If it were represented by + - - it could be confused with line 1, row 4. If it were represented by + + - it could still be confused with line 1, row 4, and if it were - - + it would not represent a correction of - + -. Thus a fourth line is clearly necessary for - + - +. This could however also be confused with line 1, row 2 (- +). The solution is to put all of the preceding categories in groups of 3's and in 2 groups of 3's or 6's.

Putting them into 3's, we get + + + for row 1, line 1, - - - for row 2, line 1, and - + - for row 3, line 1. If however we put row 4, line 1 as + - +, we will have no circular or repeatable triagrams available for the remaining four rows of line 1. By putting the information in groups of threes, both the conditions of reversal or opposites and the dialectical conditions of these opposites are met.

The solution as it is worked out in King Wen's version of the *I-Ching* is row 1, line 1 as + + + + + +; row 2, line 1 as - - - - - -; row 3, line 1, as - + - - - +; row 4, line 1 as + - - - + -; row 5, line 1 as - + - + + +; row 6, line 1 as + + + - + -; row 7, line 1 as - - - - + -; and row 8, line 1 as - + - - - -, or in the *I-Ching* symbology:

(8) (7) (6) (5) (4) (3) (2) (1)

The break where one or even three notations can be used without confusion is at line 1, row 5. In brief, line 1, rows 1 and 2, are dominated by positive/negative; line 1, rows 3 and 4 by double negative; and line 1, rows 4 and 5 by double positive. Thus rows 1 and 2 of line 1 can be represented as +, - or + + +, while rows 3 and 4 can be represented by – + –; rows 5 and 6 can be represented by + – +. Rows 7 and 8 are represented by a double negative - - –. Because there are eight rows and not four, this sequence of + + + or heaven, - + - yin or moon, + – + yang or sun, and - - - earth are insufficient to represent the logic of the system and hexagrams, groups of three notations in coupled pairs must be used. In psychological terms, the sequence becomes:

coll.	ind.	coll.	ind.	coll.	ind.	coll.	ind.	*Influences of*
8	7	6	5	4	3	2	1	*row*
(ego)	superego	(cons.)	lib.	(libido)	cons.	(superego)	ego	*Substantive category of line 1, row 1-8*

As the odd rows are dominant, and the even rows are latent, the psychological terms in the even rows may be put into parenthesis. As well the dominant odd rows may be regarded as conscious mind, while the latent even rows may be regarded as the unconscious mind, roughly equivalent to the alternating phases of conscious versus unconscious mind in the previous paradigm. Thus where libido partakes of the collective phase in row 4, for example, it is unconscious, but where it partakes of the individual phase in row 5, it is part of the conscious mind. The same is true for superego: when it is part of the collective phase in row 2, it is unconscious, but when it is part of the individual phase in row 7, it is conscious.

However, since rows 7 and 8 represent the reverse negative whereas rows 3 and 4 represent only a double negative, there is a suggestion that the roles of rows 7 and 8 may be reversed and ego may again be classified as part of the conscious mind and superego part of the unconscious mind. There is a double dialectic here, or an inversion, so that by the time we come to rows 7 and 8 in line 1 what was unconscious becomes conscious and vice versa. As another way of looking at this, since the *I-Ching* travels from the general to the specific or abstract, as it moves from right to left and from north to south, in the first four rows the alternative phases of rows represent the dichotomy of conscious and unconscious mind, but in the second four rows it represents the dichotomy between the individual and the collective mind. Thus the 'conscious' part of the 'collective' mind cannot be the superego but the ego and these two rows are seen, both by convention and logic, to be reversed. This does not happen in rows 5 and 6, however, as these rows are governed by the simple double positive. Also, for the 'conscious' part of the collective mind to be the substantive category 'consciousness' is far too abstract for this very early part of the system. Thus we have:

coll.	ind.	coll.	ind.	uncons.	cons.	unconsious	conscious	*Influence of*
8	7	6	5	4	3	2	1	*row*
superego	(ego)	(cons.)	libido	(lib.)	cons.	(superego)	ego	*Substantive Category of line 1, row 1-8*

Another more comprehensive way of governing the orientation of the rows is to realize that for the top 32 hexagrams, the *I-Ching* defines the 'individual' mind while for the bottom 32 it defines the 'collective' mind. Especially as a product of a civilization that had not yet truly acquired 4th efflorescence modes, those aspects of the individual mind seen as significant were ego/superego. Where the 'collective' mind was so closely associated with the supernatural or the 'cosmic' mind, those aspects of the collective mind deemed most significant were libido, and consciousness. Thus rows in the top half of the system were governed by the alternate phases of ego/superego, while those same rows in the bottom half were governed by alternate phases of libido/consciousness, where libido rather than consciousness was regarded as the dominant or conscious aspect of the collective mind. Thus the paradigm can be drawn more precisely:

COLLECTIVE MIND				INDIVIDUAL MIND				
superego	ego	superego	ego	superego	ego	superego	ego	
8	7	6	5	4	3	2	1	
superego	(ego)	(cons.)	lib.	(lib.)	cons.	(superego)	ego	1 2 3 4 INDIV. TO COLLECTIVE MIND
COSMIC MIND				COLLECTIVE MIND				
conscious	lib.	conscious	lib.	cons.	lib.	cons.	lib.	5 6 7 8 COLLECTIVE TO COSMIC MIND

Viewed in this way, the inversion of rows 7 and 8 in line 1 reveal a symmetry, where the governing force of ego in line 1, row 7 forces the reversal of collective superego to collective ego, and the governing force of superego in line 1, row 8 forces the reversal of collective ego to collective superego. This may also be understood in terms of placing line 1, rows 5 and 6 under Jung's category of 'extroversion' in tandem with the logical concept of 'double positive,' and line 1, rows 7 and 8 under the influence of Jung's opposing category of 'introversion,' in tandem with the logical concept of 'reverse positive' which prevails there. In Jung's terminology, where a person's libido is under the direct influence of his ego, we may say that he is 'extroverted' and when a person's superego is under the direct influence of his ego, we may say that he is 'introverted.'

The *I-Ching* is a system which from both a logical, ontological, and epistomological point of view is based on reversals, or dialectics and contradictions. Thus, while the rows from right to left, north to south, are governed by forces that go from the general to the specific, the lines are governed by forces that go from the specific to the general. In the top half of the system where the rows are governed by the individual mind or the 'general' and the lines are governed by the specific, we obtain a single contradiction or what we may call a 'double negative.' It is possible to make the general specific. However, in the bottom half we obtain a double contradiction, for how do we logically make the specific (collective mind) general? To view the collective mind as the cosmic mind is one way of doing this. In this way we enlarge the paradigm:

COLLECTIVE MIND				INDIVIDUAL MIND				
s.e.	ego	s.e.	ego	s.e.	ego	s.e.	ego	ROW-GENERAL
8	7	6	5	4	3	2	1	LINE-SPECIFIC INFLUENCE
s.e.	(ego)	(cons.)	lib.	(lib.)	cons.	(s.e.)	ego	1 consciousness
								2 ego INDIVIDUAL TO COLLECTIVE
INTROVERT		EXTROVERT						3 libido
								4 superego
cons.	lib.	cons.	lib.	cons.	lib.	cons.	lib.	ROW-SPECIFIC
								LINE-GENERAL INFLUENCE
								5 ego
								6 superego
								7 libido COLLECTIVE TO COSMIC
								8 consciousness

Since there is an additional reversal in rows 7, 8, there is also an additional reversal in the forces that govern the lines in these rows. This enlarges the paradigm again:

specific to general		8	7	6	5	4	3	2	1		*specific to general*
consciousness	8									1	consciousness
libido	7									2	ego
superego	6									3	libido
ego	5									4	superego
general to specific											*general to specific*
superego	4									5	ego
libido	3									6	superego
ego	2									7	libido
consciousness	1									8	consciousness

The first line of eight rows we have already designated as the psychic forces which also govern all the lines and rows of the paradigm in various logical orders. Logically, ontologically and epistomologically, these psychic forces may be seen as specific and abstract both, and well meet the conditions for the first line of 'specific consciousness.' Specific, individual consciousness as it moves from right to left puts ego, consciousness and libido on all rows, in decreasing orders of specificity and abstraction. Because there is a reversal of rows 7, 8, line 1, superego achieves the most general collective connotation. This reversal carries through the paradigm for all the lines 1-8 on rows 7, 8. From the point of view of specific psychic forces located in the realm of the top half of the paradigm, they also meet the conditions governing the rows in this top half for 'individual to collective mind.' From the point of view of the row alone, however, these psychic forces are under the direct influence of the general, i.e., individual to collective mind, rather than specific, i.e., collective to cosmic mind. The rows give the primary influence since they deal with the categories of knowledge or substance. The lines give a secondary influence as they are only characteristic of substance. The lines are akin to the general directional and characteristic influence which prevails on the rows as a continuum, such as the alternating phases of ego/superego in the top half of the system and libido/consciousness in the bottom half. A tertiary influence is also seen when the system is divided into groups of lines or blocks of hexagrams, such as the 'individual to collective' or 'eros' influence for the top half of the system, the 'collective to cosmic' or 'thanos' influence for the bottom half, and the 'extrovert' principle that influences rows 5, 6, line 1-4, and the 'introvert' principle that influences rows 7, 8, lines 1-4. Here the 'extrovert' and 'introvert' principles become reversed in lines 5-8. In accordance with the levels governing the total system in their various orders, we may complete the total paradigm, both ontologically and epistomologically, to obtain the master list of civilization skills which it is the function of the paradigm to detail, as can be seen on pages 281 and 282.

From both traditional methods of interpretation and from the logic of the system, the first line which delineates the psychic forces as such, as categories of knowledge, is considered both too general and too abstract to comprise specific skills of civilization. From a metaphysical point of view these categories remain in an area of pre-cognition and are essences, or vital spiritual forces that are accessible to consciousness as such but not to the action oriented psychic forces such as ego, libido, or superego. Similarly, the last line of the system by tradition is regarded somewhat outside the ken of specific skills, and when the *I-Ching* is used in application to the systems of medicine or pharmacy, the last line is regarded as particularly curative for errors in other parts of the system. Thus while the first line may be regarded as pre-cognitive, the last line may be regarded as post-cognitive. That is, these categories are also so general and so abstract at the same time, in terms of the collective to cosmic mind, that they are also in this realm seen as essences or vital social or cosmic forces. Thus as rows the categories in this final line may be placed under the influence of the psychic forces in another order as can be seen on page 283.

The existence of this order of influences for the rows of the last line suggests that under certain circumstances the entire system may not only be read from right to left and from north to south, but also from left to right and from south to north. The conditions under which this may occur has to do with the purposes of the system as a whole. Generally speaking, especially from a religious point of view, the top half of the *I-Ching* is concerned with the realm of life, while the second half is concerned with the realm of death. The division between the top 32 hexagrams and the

8	7	6	5	4	3	2	1	
(Ego)	Superego	(Conscious- ness)	Libido	(Libido)	Conscious- ness	(Superego)	Ego	1
(Kinship)	Social Organization	(Class System)	Philosophy	(Family System)	Psychology	(Logic)	Science	2
(Ritual)	Folk Religion	(Government)	Politics	(Friendship)	Ethics	(Procedure/ Routine)	Tech- nology	3
(Council)	Republic	(Intelligence)	Military	(Nationalism)	Religion/ Religious Philosophy	(Money)	Economy	4
Parliament	Democracy	Volunteer Spy	Volunteer Army	Love/ Romance	Art	Free Market	Free Enterprise	5
Emperor	Empire	Paid Spy	Mercenary Army	Fellowship	Ideology	Controlled Market	Fascism	6
Clique	Revolution	Double Agent/Plot	Army Coup	Social Psychology	Propoganda	Central Banking	Statism	7
Official Rank/Status	Bureaucracy	Education	People's Army	Ideals	Language	Taxation	Socialism	8

COLLECTIVE MIND				INDIVIDUAL MIND						
Introvert		*Extrovert*								
Superego 8	*Ego* 7	*Superego* 6	*Ego* 5	*Superego* 4	*Ego* 3	*Superego* 2	*Ego* 1			
(ego)	(superego)	(cons.)	libido	(libido)	cons.	(superego)	ego	1	*consciousness*	
(kinship)	social organi-zation	(class) system)	philo-sophy	(family system)	psychol-ogy	(logic)	science	2	*ego*	Individual To Collective
(ritual)	folk religion	(govern-ment)	politics	(friend-ship)	ethics	(proce-dure/ routine)	tech-nology	3	*libido*	
(council)	republic	(intelli-gence)	military	(nation-alism)	religion/ rel. philo-sophy	(money)	economy	4	*superego*	
COSMIC MIND				**COLLECTIVE MIND**						
Extrovert		*Introvert*								
cons.	*libido*	*cons.*	*libido*	*cons.*	*libido*	*cons.*	*libido*			
parlia-ment	demo-cracy	volunteer spy	volunteer army	love/ romance	art	free market	free enterprise	5	*ego*	
emperor	empire	paid spy	mercenary army	fellowship	ideology	controlled market	fascism	6	*superego*	Collective To Cosmic
clique	revolu-tion	double agent/ plot	army coup	social psychol-ogy	propo-ganda	central banking	statism	7	*libido*	
official rank/ status	bureauc-racy	education	peoples' army	ideals	language	taxation	socialism	8	*consciousness*	

COLLECTIVE MIND				INDIVIDUAL MIND					
Introvert		*Extrovert*							
Superego 8	*Ego* 7	*Superego* 6	*Ego* 5	*Superego* 4	*Ego* 3	*Superego* 2	*Ego* 1		*EROS*
(ego)	superego	(consciousness)	libido	(libido)	consciousness	(superego)	ego	1	*consciousness* *PRE-COGNITION*
(kinship)	social organization	(class system)	philosophy	(family system)	psychology	(logic)	science	2	*ego*
(ritual)	folk religion	(government)	politics	(friendship)	ethics	(procedure/ routine)	technology	3	*libido*
(council)	republic	(intelligence)	military	(nationalism)	religion/rel. philosophy	(money)	economy	4	*superego*
COSMIC MIND				COLLECTIVE MIND					
Extrovert		*Introvert*							
cons. 8	*libido* 7	*cons.* 6	*libido* 5	*cons.* 4	*libido* 3	*cons.* 2	*libido* 1		*THANOS* *LINE OF DEATH*
parliament	democracy	volunteer spy	volunteer army	love/ romance	art	free market	free enterprise	5	*ego*
emperor	empire	paid spy	mercenary army	fellowship	ideology	controlled market	fascism	6	*superego*
clique	revolution	double agent/plot	army coup	social psychology	propoganda	central banking	statism	7	*libido*
official rank/status	bureaucracy	education	peoples' army	ideals	language	taxation	socialism	8	*POST-COGNITION* *consciousness* *LINE OF LIFE*
ego	*superego*	*consciousness*	*libido*	*ego*	*consciousness*	*libido*	*superego*		*INFLUENCES OF THE LINE OF LIFE*

bottom 32 is often regarded as the 'line of death.' This is the division between the 'upper world' and the 'under world.' The last line however is often regarded as the 'line of life,' and is the means by which, according to the purposes of the system, reincarnation could occur. Thus the skills of civilization delineated in the bottom half of the system have a double connotation. For one thing, they are regarded as the skills of civilization particularly important in the development of the collective mind or collective modes of behaviour, and they presuppose a human collectivity of greater sophistication and organization than that presupposed in the top half of the *I-Ching*. But they are also regarded as the skills of civilization necessary for life in the 'under world' as opposed to the 'upper world.'

In Chinese myths and practices relating to the underworld most of these civilization skills are clearly evident. Like the Egyptians, ancient Chinese burial sites reveal, through the actual dead personages or their model representations or representatives, the civilization skills associated with the Emperor, his clique, his spies, mercenaries, plots, intrigues, fellowship, love, art, ideology, propoganda and his market systems. The theory apparently was that if all of these civilization skills were appropriately represented, the last line of the system would be made existent as the completion of the various dialectics by rows evolved. A kind of collective/cosmic 'kingdom of heaven' could be achieved where the most general principles of collective and cosmic modes would prevail, and this would provide the 'line of life' by which reincarnation could be achieved. These fair principles of the last line of the system, however, were seen as not the means but the ends of collective modes, which explains how such social brutalities and excesses could be logically justified, since in the various categories which lead to the 'kingdom of heaven' by dialectical evolution in the rows, all of the social and political brutalities and excesses characteristic of these ancient civilizations must first be present.

Even from a superficial point of view the mathematical and geometric possibilities of this system are almost immediately evident. The influences governing rows and lines both represent straight lines in a perpendicular relationship to each other. Since the first line of pre-cognition and the last line of post-cognition are more akin to the rows and lines of influence than to the specific categories of substantive knowledge which the system delineates, from a logical point of view these two lines, the first and the last, are redundant with the rows and lines of governing influence. Thus, in the top half of the system there are eight rows of 3 categories of substantive knowledge (lines 2-4) whose substance is produced by the lines at right angles to each other which act as governing principles, and the same is true in the bottom half of the system (lines 5-7). This total of 16 rows of 3 categories of knowledge directly represents the hypotenuses of 16 right angled triangles.

For a perfect geometrical representation of the conjunction of any two 'hypotenuses' or the conjunction on two rows on lines 2-4, it is seen that the top and bottom resultant hypotenuse of ego/libido/superego and superego/libido/ego, represents the direction in which the line influences is read. Since the odd rows in the top half of the top half of the system represent conscious mind and the even represent unconscious mind, influences from the bottom half of the system makes itself felt in the even rows, as opposed to the odd, as the bottom half as generally collective and cosmic is also unconscious. Thus there is an influence on the even rows even in the top half of the system which represents the 'way of death' or 'thanos,' while the odd rows alone—with the exception of the reversal that occurs in rows 7 and 8—are only in the 'way of life' or 'eros.'

In the bottom half of the system where more lines of influence exist, and where there is no distinction between conscious and unconscious phases in a mind that is anyhow unconscious, this geometrical representation unfolds like a cube. (See pages 286, 287). For rows 1, 2, lines 5-7, the 4 lines that form the perpendiculars of the cube are represented by the influences on the rows read from right to left and from north to south and in reverse with redundant categories eliminated. The four lines that form the parallel two faces joined by the four perpendicular lines are represented by the 4 lines of influence from the top half of the system and from the 4 lines from the bottom half also read from north to south, and then from south to north in both cases. The four diagonals of the cube are then the categories in rows 1, 2, lines 5-7 read first from north to south and then south to north in both cases.

It can be seen that the point of intersection of one side of the cube can be represented by 'controlled market' and the point

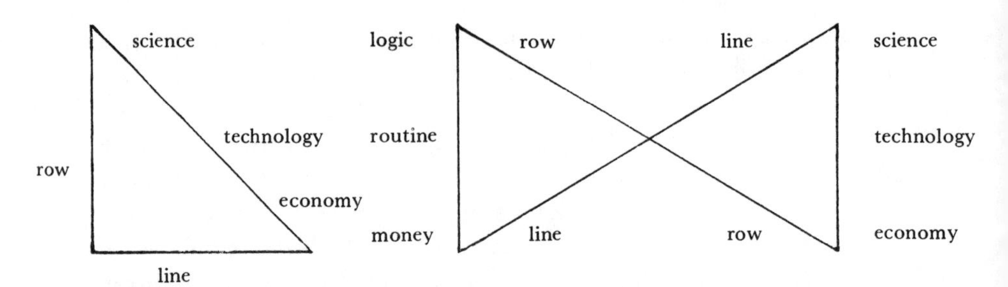
science
technology
row
economy
line
logic
routine
money
row
line
line
row
science
technology
economy

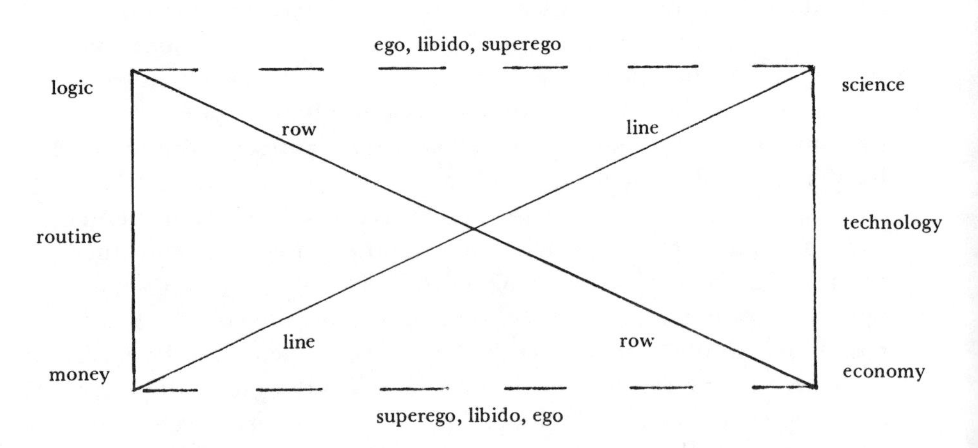
ego, libido, superego
logic
science
row
line
routine
technology
line
row
money
economy
superego, libido, ego

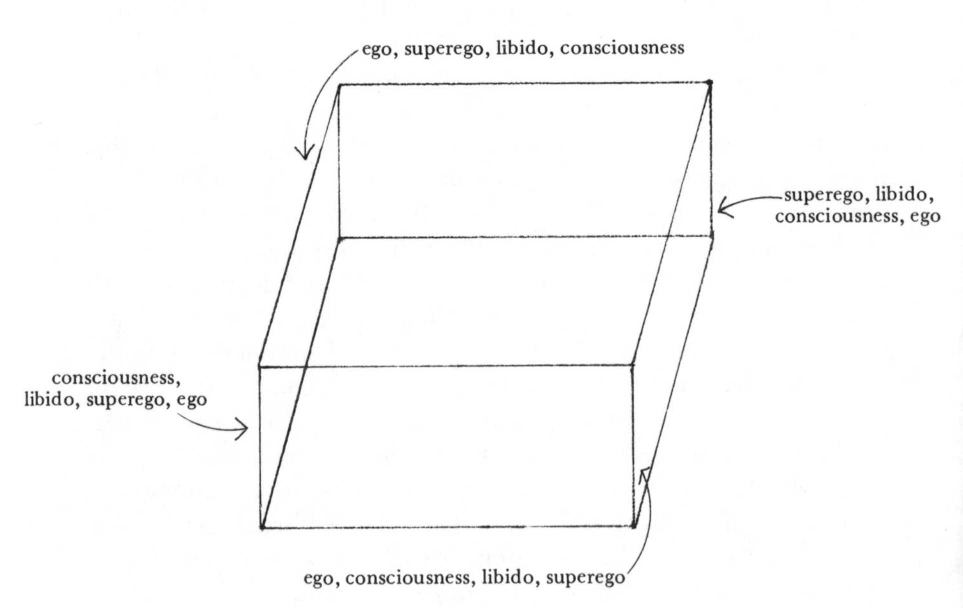
ego, superego, libido, consciousness
superego, libido,
consciousness, ego
consciousness,
libido, superego, ego
ego, consciousness, libido, superego

of intersection on the other side of the cube by 'fascism.' Since the same lines and rows of influence prevail over all of the 24 categories of information in eight sequences of 3's for the diagonals of right angled triangles, it is possible to contain all eight of these sequences on a single geometrical figure. The middle line of these sequences of 3's, on the bottom half of the system (for rows 1-8, rows 5-7) line 6 would represent the intersection of diagonals for 8 diagonals at a different angle again representing 'fascism,' 'controlled market,' 'ideology,' 'fellowship,' 'mercenary army,' 'paid spy,' 'empire' and 'Emperor.' This geometrical figure would be a four-sided pyramid with resultant categories represented as laterals of the four sides of the pyramids (see page 288). A cube of six sides with a pyramid projected on each side would account for all the possibilities of line 6, eight categories that must be arranged in consecutive groups of 2's along line 6.

The fascination of the ancient Egyptians with pyramids and the use of four-sided pyramids on the tombs in which the body of the Pharaoh was placed to facilitate his reincarnation can be seen as more than accidental when we realize the pyramidic structures that evolve directly from the logical system of the 'master code.' From the logic of the system in the context of its original purpose,

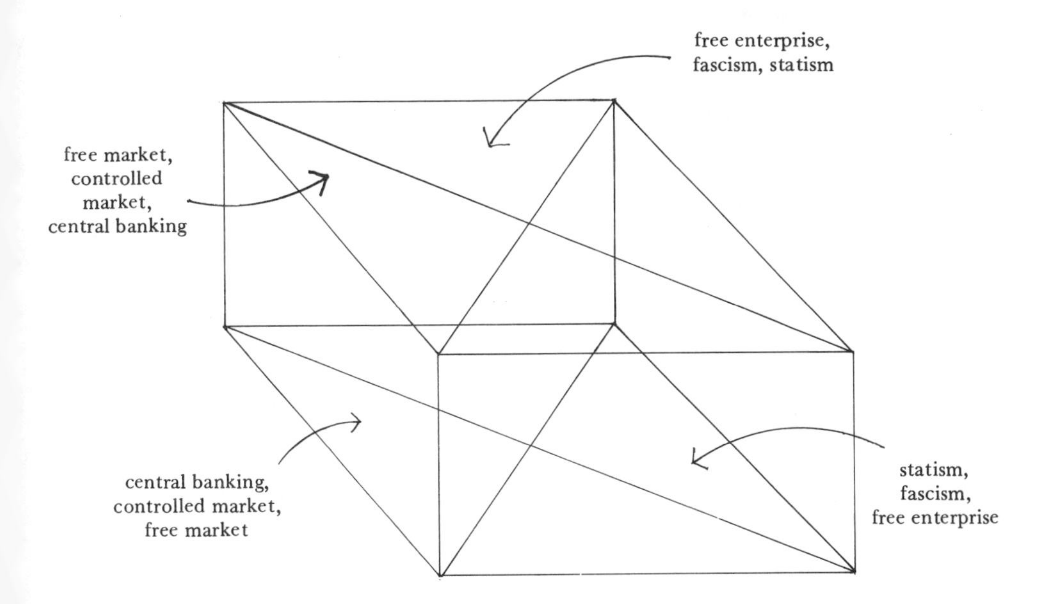

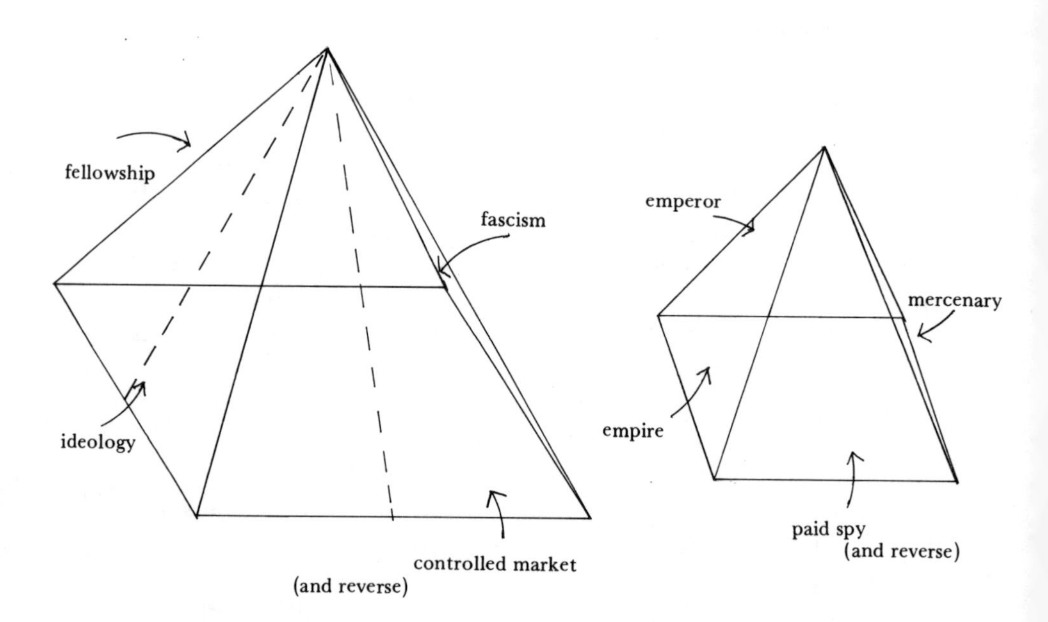
fellowship
fascism
ideology
controlled market
(and reverse)
emperor
mercenary
empire
paid spy
(and reverse)

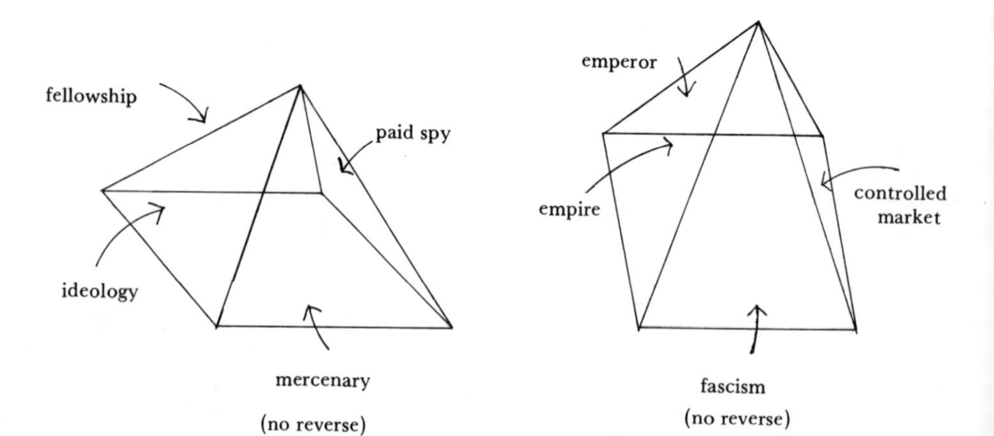
fellowship
paid spy
ideology
mercenary
(no reverse)
emperor
empire
controlled
market
fascism
(no reverse)

no doubt the most important sequence for a pyramidic construction was the sequence, 'emperor'/'empire'/'controlled market'/'facism.' This sequence is not amenable to reversal because the 'emperor'/ 'empire' conjunction comes under the influence of rows 8, 7 which have already been reversed. In conjunction with 'controlled market'/'fascism' a circular movement is indicated as the rows involved, 7, 8 and 1, 2 are at opposite ends of the system and can only be joined by a circular movement from row 8 to row 1.

In this way the abstract and unconscious influence of superego and ego indicated by the forces that govern rows 7, 8 become identified totally with the general and conscious influence of superego and ego that govern rows 1, 2. In this way the individual, collective, and cosmic ego become entirely identified and indistinguishable, which was indeed one of the distinguishing characteristics of Pharaonic rule. In a very graphic way the construction of pyramids whose pyramidic faces extended upwards from the face of the earth represented the re-emergence of the Pharaoh from the 'underworld' to the 'upper world,' or his physical reincarnation. In this case the influence of 'consciousness' is minimal, having an influence only on rows 7, 8, 1, 2 through the middle of the system at the 'line of death.' The minimal influence of 'consciousness' on the upper face of the pyramid may have been compensated for by building the base of the tomb, within the earth, in the form of another pyramid representing rows 3, 4, 5, 6 which represented the sequence on the laterals of this 'underworld' pyramid, 'ideology'/'fellowship'/'mercenary army'/'paid spy.' In these rows 'consciousness' as a governing force is present six times, whereas in rows 7, 8, 1, 2 it is present only twice and in positions of minimal importance. This may in part account for the necessity, in Pharaonic tombs, of burying, either in person, or in representation, the entourage of the Pharaoh and displaying the hieroglyphic codes which detailed the ideology of the ancient Egyptian empire.

In the last line of the system, the 'line of life,' the categories of knowledge are closely identified with the influences that govern these categories by rows. This line of influence permits the entire system to be read from south to north as well as from north to south. From a geometrical point of view it represents another straight line at a different angle from the other two straight lines at 45 degree angles to each other representing the influence of rows and the influence of lines. It permits the entire system to become three dimensional, and it allows that which is collective and cosmic to become individual again.

Since the line of influence governing the categories of knowledge on the 8th line as rows were so closely identified with the categories themselves, a society governed by this system would not only see the achievements of these categories as the end accomplished through a variety of means, many of which were not particularly pleasant, but also would see the achievement of these categories in the existent society in some form as a mark that the entire process of the system was being followed and was successful. Something like the concept of the 'elect' in 17th century Protestantism, the existence of these categories in societies governed by Pharaonic rule and following the processes of the 'master code' was visible proof of the cosmic and supernatural efficacy of the system. The existence of the categories 'official rank,'/'bureaucracy'/ 'socialism'/'taxation'/'education'/'peoples army'/'language'/'ideals' in the society were not valued as ends in themselves, nor were they highly institutionalized. Rather their existence in vague but identifiable form had the aspect of 'becoming' and were evidence of the 'heaven on earth' promised by the physical reincarnation of the God-King or Pharaoh.

Chapter XVIII

THE PERSISTENT INFLUENCE OF THE *I-CHING* ON BOTH ANCIENT AND MODERN CHINESE CIVILIZATION

It is extraordinary when we look at the work of Confucius, his contemporaries, and his followers to realize how central an influence this system of the 'master code' or *I-Ching* was to the post-historical Chinese tradition. The commentaries of Confucius, his contemporaries, and his innumerable followers throughout the 2500 years of post-historical Chinese civilization make up almost the entire bulk of official Chinese literature, history and philosophy. The original 'master code' of the Egyptians had been resuscitated by King Wen, and Confucius after him, in order to revise it in terms of an ego-negating and libido-abdicating religious philosophy. While all the categories of knowledge in the 'master code' were acknowledged by Confucius, his contemporaries, and his followers, those categories which conflicted with the ego-negating and libido-abdicating purpose of the religious philosophy they espoused were singled out only to be ignored. Except for the last line of the system, Confucius himself ignored all of the categories of knowledge in the bottom half of the *I-Ching*, although some of his contemporaries and his followers did emphasize some of the categories or processes strictly in isolation from other categories in this part of the system.

As a religious philosopher, Confucius was most noteworthy in that he placed almost no importance on supernatural religion *per se*. The bottom half of the system which deals with the 'collective to cosmic' mind inherently contains all of the aspects of supernatural religions and is directly identified with life after death or the 'underworld.' However, since the social and political excesses of Pharaonic rule were so closely identified with this part of the system, Confucius, who nonetheless had a powerful religious sensibility, preferred to ignore almost in its entirety this aspect of the system and all of the skills of civilization associated with it.

Confucius was far more comfortable with the top half of the system except where the categories came into direct conflict with his ego-negating, libido-abdicating values, and all of the categories which Confucius promulgated as precepts of his teachings occurred in the top half of the system such as 'procedure/routine,' 'logic,' 'psychology,' 'ethics,' 'family system,' 'friendship,' 'philosophy,' 'politics,' 'class system,' 'government,' 'social organization,' 'folk religion,' 'kinship' and 'ritual.'

With the exception of row 1, which falls directly under the influence of 'ego' twice by row and 'ego' and 'libido' by line for 'science' and 'technology' which he ignored, all of the categories he valued occur in lines 2 and 3. Line 4 implied the dialectic completion of lines 2 and 3, for the conscious or individual to collective mind. Confucius and his contemporaries wished to eliminate this sense of dialectical completion from their religious philosophies since this was the basis on which the surrogate Pharaonic structure they wished to totally eradicate had been built and sustained. They had hoped to resuscitate the entire 'master code' in order to cull from it those skills of civilization that were compatible with their ego-negating and libido-abdicating orientations and then to eradicate the system as a whole from conscious, historical memory. The conundrum in which they then found themselves helps to explain many of the contradictions of Confucianism and the sense of incompleteness which characterizes so much of Confucianism as both a religious, psychological, and social philosophy and as an abstract system of thought.

Thus, while all of the categories of knowledge of lines 2 and 3 with the exception of science and technology were openly acknowledged as important by Confucius, some were more highly valued than others depending on their exact position in the system. For example, although philosophy and politics were acknowledged by Confucius as extremely important, they were placed under severe constraints of moral thought and behaviour in attempts to mitigate by especially stringent ego-negating and libido-abdicating imperatives their primary influences of 'ego'/'libido' (by row) and 'ego' (by line) in the case of 'philosophy' and 'ego'/'libido' (by row) and 'libido' (by row) in the case of 'politics.' Confucius tended to intertwine 'philosophy' as a category with its complement, the 'class system,' since the latter fell under the more auspicious influences of 'superego'/'consciousness' by row.

Class system was not seen in terms of institutionalized slavery of any kind but was modeled on the family system, a hierarchy determined by order of birth and sex, and parent/child relationships and governed in the most stringent way by the ego-negating and libido-abdicating moral imperatives. Modeled on the family system, the class system was composed of family units in extended familial hierarchies which could be superceded only by official status in the bureaucracy. The relationship that members of family units had with bureaucratic officials lacked any connotations of subservience or even obedience, and comprised an almost religious respect which befitted the position of official status/bureaucracy in the cosmic and curative position of the system as a whole. Significantly, the ordinary man or citizen had no direct relationship with the king or Emperor, whose exact role is extremely vague in Confucianist thought and lacks all connotations of direct power over his subjects as evident from the absence of direct relation of subject to king or emperor. In Confucian thought, only the official in the bureaucratic structure had a direct relationship to the king or emperor, although again the relationship of the official to the king was characterized by loyalty rather than subservience or even obedience.

As befitted their relative positions in the system, especially in the context of the revision of the system in terms of the ego-negating and libido-abdicating orientations of Confucianism, the official in Chinese society was always more important in the Chinese political and social system than the king or emperor himself. Only in those periods, which have been brief by comparison with the duration of post-historical Chinese civilization, when there was an attempt to overthrow Confucianism as a mode of thought, such as during the time of the Chin dynasty some 2200 years ago, or recently, during the time of Mao's reign, has the relative importance of these relationships been reversed. Thus the emperor of Chin, whom Mao was later to use in some part as a model for his own efforts in overthrowing Confucianist thought in China, emphasized certain isolated categories of the system, such as economy, money, fascism, controlled market, statism, central banking, empire and emperor, that fell under the strong influences of 'ego' and 'libido' and were largely in the 'collective to cosmic' realms rather than in the 'individual to collective' realms. The emperor of Chin however was highly constrained in his efforts to overthrow Confucianism by the fact that even his most sympathetic theoreticians such as the 'legalist' Han Fei-Tzu remained under the direct influence of Confucianist tradition. The secrecy which had surrounded the resuscitation of the 'master code' by King Wen and Confucius had led to a deliberate obfuscation of the keys which allowed the system to be understood in its entirety.

From the advantageous position of historical hindsight, we can see that this obfuscation was accomplished by neglecting to properly translate the first line of the *I-Ching*, which in turn made it impossible to precisely comprehend the influences governing the rows and lines of the system. Although Confucius alludes to psychology and indeed all of his work is permeated with an intriguing psychological sensitivity and insight, he stops short of developing a psychological system or relating the central triagrams to psychic forces directly, which would have made the 'master code' accessible in its entirety to later generations. In addition, so thoroughly had all information about the way of life under surrogate Pharaonic rule been eliminated from Chinese written records by the time of Confucius that even such a dedicated fascist as Han-Fei Tzu could hardly have visualized the mechanisms by which the ancient Chinese despots held total sway over their subjects.

By modern times however enough information was available to Marxist theoreticians in China loyal to Mao to resuscitate the *I-Ching* and utilize it for the purposes of Marxist revolution. Under the direct influence of Marxism and in the context of their Westernized education, these theoreticians were able to use borrowed objectification of phenomena modes for what they deemed to be the purposes of establishing a Communist state under Chinese conditions. Due to the adverse effect that Stalinism had on them, however, as a doctrine, these theoreticians were inhibited from making the necessary direct linkages between individual psychic modes and the eight hexagrams of the first line of the system which would have enabled them to translate the system in its entirety. Under the influence of Stalinist doctrine they directly identified the individual psychic modes with that of the collective, and in seeking to obliterate individual psychic modes *per se* they concentrated on the bottom half of the system, the realm which Confucius, his contemporaries, and his followers had for the most part ignored.

As Maoist theoreticians were unable, due to the doctrinare atmosphere in which they lived and worked, to grasp the system in its entirety, they were also unable to grasp the purposes for which the system had been devised in the first place. Under the influence of Stalinism they had denied the integrity of religion as a valid social force, so that the supernatural and cosmic aspects of the lower system eluded them, believing as they did that it applied solely to collective modes rather than to a mode that was both collective and cosmic at the same time. Under the influence of Stalinism in Russia, there had been an attempt to develop 4th efflorescence modes by directly substituting 2nd efflorescence modes for 4th in the guise of a cult hero, Stalin.

In Nazi Germany this substitution had been clearer since institutionalization of 3rd efflorescence modes was less well entrenched than in Russia where the population was a Nordic and Asiatic mix. The universal moral values contained in Communism as an ideology was transferred in Russia through these well-entrenched 3rd efflorescence modes. In both the cases of Stalinist Russia and Nazi Germany, however, the attempt to develop 4th efflorescence modes by directly substituting 2nd efflorescence modes in the guise of a cult-leader or Führer as a model of both personal and collective behaviour was motivated by pragmatic considerations and could to a large extent even be considered instinctual. These pragmatic and non-cognitive modes of approaching culture were characteristic of Western civilization as it had developed on the Grecian model. Such was not the case in China. Once the doctrine of Stalinism was established in China, the Chinese were inhibited by virtue of their cultural modes from proceeding on an entirely pragmatic and instinctual basis to fulfill the logical precepts of Marxist revolution.

The breech between Maoism and Western schools of Communism was born of the Chinese refusal to accept the absolutism of the Marxist doctrine, which in Chinese terms meant accepting a total objectification of phenomena which was alien to their cultural modes of both thought and behaviour. While it was logically correct to propose that Communism was an absolute and was constrained by particular historical conditions, mired as they were in the Stalinist doctrine, the Chinese were unable to look at the Communist system with any real objectivity or to observe in terms of any real dedication to the truth its logical validity. In this they abdicated the tools of analysis which were Chinese and native to them and the tools of analysis which were Western and accessible to them. Instead, they chose to select that part of the Stalinist doctrine that could be most easily applied within the framework of the historical tradition which was their legacy. Thus, although they attempted to create a uniquely Chinese cultural revolution, the Maoists succeeded only in obliterating that part of their heritage which was Confucianist and maintained the general framework out of which Confucius himself had deliberately selected those aspects of the civilization skills which were compatible with his own ego-negating, and libido-abdicating orientations. However to attribute the ravages that Maoist thought inflicted on Chinese society to logic is no more devastating than the attribution of the ravages caused by Stalinism and Hitlerism in Western society to unbridled pragmatism and psychic urges. The excesses of logic on the one hand and pragmatism and psychic urges on the other merely reflect the modes of these differing societies, which, when exercised under constraint and caution had provided uniquely adaptive cultures within the framework of differing historical circumstances.

The theories of Maosim, inspired by Stalinism and buttressed by a re-examination of their own native 'master code' in the light of Marxist theory, held that the stages by which Communism were to be achieved could be accelerated to span the single lifetime of the cult-hero, Mao, in whose collective person the 'heaven on earth' of proletarian dictatorship could be embodied. In the span of Mao's lifetime and rule, early stages of this dialectical process were allowed to occur while he was still leader of the guerilla army in Yunnan and during the first years after his reign over all of China. Thus, in this time period all the categories of line 5 of the *I-Ching* system were permitted full expression, i.e., 'parliament,' 'democracy,' 'volunteer spy,' 'volunteer army,' 'love/romance,' 'art,' 'free market' and 'free enterprise.' In the second period of his reign when these categories had, in the eyes of Maoist theoreticians, run their course in terms of having accomplished the first stage of the Communist dialectical process, they were disallowed, at which time, for a period of about ten years during which time the Korean war occurred, all of the categories of the 6th line of the *I-Ching* were allowed full expression. It was during this period that Maoists irrevocably broke with the Russian Communists to whom it had finally become apparent that the Communist revolution was not to be of the standard type they demanded in China. Mao had to all intents and purposes declared himself 'emperor' of a revitalized Chinese 'empire' which was upheld by a stringent Maoist 'ideology.' 'Espionage' was blatantly bought and sold, and forms of 'mercenary armies,' whose pay was recouped by the 'fascist' state, were dispatched to all parts of the globe, including Korea, and whose cadres sacrificed their lives by the thousands on the basis of 'fellowship.' At home the 'market' became 'controlled.'

About 1963 the cultural revolution was launched at which time the categories of line 6 of the *I-Ching* system became disallowed while the categories of line 7 were permitted full expression. This was to be a society of 'endless revolution' run by infamous 'cliques' such as the 'gang of four,' where 'double agents' or internal espionage was used against the enemies of the revolutionary state at home and abroad. This period was also characterized by 'military coups' which however proved abortive such as in the case of Lin-Piao. The self-criticism sessions of this period which were universally practiced by the masses exhibited a sophisticated science of 'social psychology,' which replaced family, friendship and fellowship among the cadres and which was indiscriminately lent abroad for the purposes of recruiting sympathizers and informants in unlikely left and right regimes alike. 'Centralism' was established as an economic mode, under the aegis of 'statism' and with an organized 'central banking' system.

With the death of Mao, tremendous uncertainty exists as to whether or not to attempt to implement the 8th line of the system, where the categories of knowledge promise the 'heaven on earth' of collective utopia in the *I-Ching* system, i.e., those of 'official rank or status,' 'bureaucratic society,' 'education,' 'peoples' army,' 'language,' 'ideals,' fiscal systems of 'taxation,' and 'socialism.' The ravages to Chinese society, the political and social excesses they caused, have served to call into question the credibility of the entire dialectical process by which this ultimate synthesis was to be achieved, and hence the ultimate synthesis itself. It is true that since the death of Mao, 'official status and rank' has been reinstated and there has been an attempt to establish a 'bureaucratic society.' There have been attempts at 'language' reform, resinstitution of 'ideals' as opposed to ideology and propoganda, and an attempt to institute proper 'education' on a mass scale. However, the concept of a 'peoples' army' which was very popular in ruling circles during Mao's last days and shortly thereafter may have fallen into some disrepute, and there appears to be considerable anxiety as to exactly what form the economic and monetary aspects of the society should take, although indeed 'socialism' as opposed to both capitalism and Communism seems to be the prevailing mode. Also, with differential earnings now allowed for workers, a system of 'taxation' has been instituted, although it is generally acknowledged that it will be some time indeed before 'taxation' rather than 'central or collective banking' will be able to bear the fiscal burdens of the state. With the possible exception of the 'peoples' army' all of the categories of knowledge in the 8th line of the *I-Ching* are more or less compatible with an ideal mode of 4th efflorescence society, such as has been achieved in certain small European countries.

To affirm that the attainment of these categories is the ultimate end of a historical process is to affirm, on the one hand, a kind of 'extended Marxism.' Marx believed that the dialectical process as developed by the Greeks and refined by Hegel could be applied to material reality. Taking a general evolutionary perspective, Marx found the first historical 'thesis' to be the development of the slave-state of pre-historical times. This 'thesis' developed in time into the 'anti-thesis' of bourgeois capitalistic society, where the bourgeois owned and controlled all of the economic means of production. A new 'thesis' arose when the workers developed as a self-conscious class, and the ultimate historical 'synthesis' was to be achieved at the point when the workers obtained, through class struggle, the economic means of production in their own hands. Marx envisioned the means by which all the skills of civilization were produced to be primarily a production of economic evolution, and he saw economics as the first cause from which all the other skills of civilization necessarily followed. Thus, forms of government, types of culture, religions, and prevailing social ideals were a product of the particular economic form that existed at any point in time of historical evolution.

The implications of Marx's theories is that historical evolution was a natural process, at least up until the time when the workers emerged as a self-conscious class. After this point, natural evolution was no longer the mode by which historical development occurred; rather evolution became man-made, as the workers, with the guidance of the educated and ideologically-oriented Communist party members, inspired class struggle and used every means at their disposal to attain the desired end where the economic means of production would be squarely placed in the hands of the working class. It was the man-made aspect of this last stage of the historical dialectic which was found so appealing by Maoist theoreticians who believed with the aid of the ancient 'master code' that they could more exactly systematize it. According to the *I-Ching* the economic forms were not the first cause of all the skills of civilization but were a separate dialectic in and of themselves, specific categories of civilization skills among a number of others. To ensure the success of the ultimate historical dialectic, all the specific skills of civilization, as dialectics, must by self-conscious means be allowed to run their course. However, from a logical point of view if the *I-Ching* system were to be affirmed as a precise model of social reality, the fallacy of Maoist theoreticians was to totally disregard the top half of the system and accept only the bottom half.

An even more profound criticism of this method of 'extended Marxism' in the application of the *I-Ching* to social reality, is that there is no indication in the system itself that these categories of knowledge as civilization skills are in any way either inclusive or exclusive of each other, neither is there any indication that the dialectic run on each row of categories is achieved through class struggle or any other kind of struggle, man-made or otherwise. The system is intrinsically static, given the original purposes for which it was constructed. On the one hand it was a means of organizing in tabular form all the skills of civilization born out of peak 3rd efflorescence society. On the other hand it was oriented in a way that buttressed and fed the ego-recognition and libido aggression of the Pharaonic rule for whom it was aimed to deify. Nonetheless, Maoist theoreticians were impelled under Stalinist doctrine to achieve a Communist state in China, and appalled as they were by the failure of China to develop more than the embryonic forms of 4th efflorescence society for which a Communist state was seen as only the latest stage, Maoist theoreticians resorted to the more precise dialectical system of their own native culture, the *I-Ching*. Their interpretation of this system, like Confucius before them whom they so vehemently and violently disowned, was biased. Whereas Confucius' interpretation of the system had been influenced by his ego-negating and libido-abdicating orientations, the interpretation of *I-Ching* by Maoist theoreticians were influenced by Stalinist doctrine.

Just as Confucius' personal orientations had permitted him to infuse the system at every point with philosophical, and moral imperatives that were entirely external to the system, so the personal orientations of Maoist theoreticians, which were Stalinist, led them to infuse the system at every point with the sense of class struggle that was entirely external to the system. Just as Confucius' personal orientations led him to entirely ignore the bottom half of the *I-Ching*, so the personal orientations of Maoist theoreticians led them to entirely ignore the top half. The way in which Confucius infused moral imperatives into those parts of the system he found compatible with his own personal orientations were no less artificial than the way Maoist theoreticians infused imperatives of class struggle and conflict in those parts of the system which they found compatible with their own Stalinist orientations. It is extremely interesting however that those categories that Maoist theoreticians infused with class struggle and conflict in a most deliberate, amoral way were seen as temporally limited. Once the dialectics of lines 5, 6, and 7 had run their course, according to the Maoist theoreticians, class struggle and conflict would come to a final end in the realization of the ultimate historical dialectic of line 8. The theory was that in the brief span of Mao's reign, almost, as a matter of fact, about the same time that the Emperor of Chin reigned, about 30 years, the social and political excesses that arose from class conflict and struggle were permissible. After this brief period of time, however, when the ultimate historical dialectic of line 8 was achieved, class conflict and struggle would no longer exist and social and political excesses of every kind would be disallowed.

According to the ancient system the 8th line of the *I-Ching* was the 'line of life' and represented a particularly moral and curative dialectic that led directly to the top half of the system. As the top half of the system deals with the individual-collective mind it lent itself very well to the ego-negating and libido-abdicating orientations of Confucius, who, from all the categories in the bottom half of the system, found only the 8th line to be compatible with his own religious philosophy of ego-negation and libido-abdication. China today reaffirms itself as a socialist state, but it has denounced the man-made historical process by which this state was achieved during the reign of Mao. Those categories of knowledge which were particularly instrumental in leading to social and political excesses during this time, such as 'Emperor,' 'empire,' 'clique,' 'ideology,' 'propoganda,' 'social psychology,' and 'fascism' have been singled out and denounced. Whether this means that China will in the future abandon Marxism, or in particular Stalinism, in terms of its dialectical theories of class struggle and historical evolution remains to be seen. Undoubtedly however the ego-negating and libido-abdicating orientations of Chinese civilization are deep-rooted and do not derive directly from Confucius, who only crystallized these orientations, but from the *summa* of a vast historical experience and evolutionary process. With these deep-rooted orientations it is more likely than not that China will remain, in spite of its conscious socialist goals, more in the realm of the late stages of the 3rd efflorescence than in the 4th, or at least in a realm which allows it to straddle both.

In accepting this 'master code' as the one and only logically verifiable system upon which a civilization could be founded, Confucius was forced into a series of contradictions, many of which still have yet to be unravelled. On the basis of his own ego-negating and libido-abdicating orientations he, his contemporaries, and his followers were forced to emphasize certain categories of knowledge or skills of civilization and deemphasize others in a way that seems random and artificial without an understanding of the 'master code' in its entirety. Due to the overwhelming influence of ego-recognition and libido-aggression on 'science' and 'technology' in the system, these civilization skills were ignored much to the detriment of Chinese civilization. 'Economy' was acknowledged, and indeed the Chinese independently developed both 'money' and 'banking,' but out of the context of the 'science' and 'technology' in the system that made 'economy' possible, it languished, and merchants were inevitably seen as holding the lowest rank in the social order. Although 'logic' and 'procedure/routine' were allowed, without their complements in 'science' and 'technology,' their mechanisms remained obscured and were not widely disseminated. Like 'religious philosophy,' 'nationalism' was acknowledged but because of its direct connection with the 'collective-cosmic' part of the system which Confucianism abhorred it was never extended in any systematic way. Due to the fact that the row sequence 'philosophy'/'politics'/'military' fell under the pejorative influences of 'ego'/'libido' all of these categories were put under severe restraints.

'Philosophy' was never allowed to reach a high level of abstraction, and was used as a mediator between the 'family system' and the 'class system.' Like 'politics,' it contained artificial moral imperatives that resulted from the ego-negating and libido-abdicating orientations of Confucianism. 'Politics' was intertwined with moral 'government,' and based on the moralistic model of 'friendship.' With the inhibition of 'politics' and 'philosophy,' 'republic' as a category of knowledge or civilization skill seemed irrelevant, and decision by 'council' seemed unnecessary. The 'military' was acknowledged but due to the pejorative influence of 'ego' and 'libido' on it and its direct connection with the collective aspect of the system, it received little emphasis in Chinese civilization, and men of the military were designated a social status only next to that of the merchant. Although 'intelligence' was never directly taken up by Confucius himself, due to both its position on the system and the pejorative connotations of associated skills such as the 'military,' 'politics,' and 'philosophy,' it was of extraordinary importance in post-Confucian thought.

One of the greatest classics of Chinese civilization, *The Art of War*, by Sun Tzu, a contemporary of Confucius, deals in great detail with the sequence 'intelligence'/'volunteer spy'/'paid spy'/and 'double agent.' The aim of the *Art of War* is to substitute wherever possible the skills of 'intelligence' for the skills of direct military action, and it examines with exquisite precision all the ways that the fate of the nation may be ensured in times of war by a proper exercise of these skills.

'Social organization' was ignored by Confucius no doubt because of its double 'superego' influence which so directly associated it with the collective rather than the individual mind. The abdication of 'social organization' as a civilization skill was compensated for by an intense emphasis on its complement, 'kinship.' In the same row with double 'superego' influence, i.e., row 7, which was reversed to row 8 position, 'folk religion' received little sympathy from Confucius, and its abdication too was compensated for an intense emphasis in Confucianism on its complement, 'ritual.' Of all the categories ignored or abdicated by Confucianism, none was taken up with more success by competing schools of Taoist thought than that of 'folk religion.' The modes of 'folk religion' which the Chinese masses had inherited from the times of surrogate Pharaonic rule have stubbornly persisted in China until modern times. As the masses' version of this 'master code,' modes of folk religion provided a continuing wealth of knowledge and information through which an understanding of the 'master code' could be retained and transmitted in spite of the rigorous ethical codes and official dogmas of the Confucian state which inhibited any formal explication of it.

It was no doubt the severe restrictions that Confucianism as a whole placed on the full and natural flowering of all civilization skills during the 2500 years of its supremacy in China that resulted in the upheavals in China during modern times and ultimately the Communist revolution there. Lacking adequate sustanence from the colonial powers in China to allow a gradual reform or a more rational revolution under the leadership of Sun Yat Sen, the Chinese fell into the seductive lure of Stalinism and attempted to undo by force the restrictions that 2500 years of Confucian rule had placed on their natural evolutionary, and historical development.

Thus, in one way or another all modern civilizations live under the long shadow of the ancient Egyptian empire. The emergence of Egyptian civilization was an accident of human history and evolution, while everything that subsequently followed had more the sense of logical necessity than accident. In its importance to the human race, the accidental emergence of Egyptian civilization was as important as the emergence of the first mammal in an upright stance. Once upright stance was achieved by mammalian species, it seemed certain, from the perspective of evolutionary hindsight, that ego-recognition and superego entrenchment would inevitably follow, given time of sufficient duration. In other words once first efflorescence modes were established, it became a logical necessity that 2nd and 3rd efflorescence modes would follow. While 2nd efflorescence modes were relatively general, however, the way in which 3rd efflorescence modes were entrenched were relatively specific, depending on a number of external factors and the often accidental way in which superego entrenchment became attached to them.

Egyptian civilization resulted in a re-generalization of 3rd efflorescence modes, making them as universally appliable as 1st and 2nd efflorescence modes had been in their time. Third efflorescence modes, it might be said, portended the first significant specifiable differentiation in humanoid efflorescence. Had it not been for the generalizing effects of the Egyptian empire, civilization as we know it might still not exist in the world, or might have an entirely different configuration, making it to all intents and purposes unrecognizable to the modes of civilization that have developed on this planet. The effect of the generalizing aspects of the Egyptian empire have proven so pervasive that no culture having developed independent 3rd efflorescence modes has been able by this point in time to escape its direct influence. From the generalization of 3rd efflorescence modes that Egyptian civilization established, all the modern civilizations, whether in the late stages of the 3rd efflorescence or in the 4th, or somewhere between the two, were derived. Historical accident, as well as the now less important external factors of environment, played the central role in determining the specific configurations of all civilizations that had their roots in the Egyptian empire.

Existing cultural diversity can be explained by the way the cultural and racial ancestors of the various modern civilizations reacted and interacted with Egyptian civilization. With the sole exception of those heirs to the Grecian model of civilization in the so-called 'developed' world, all of the extant civilizations in modern times are the heirs of civilizations who enjoyed prolonged and direct experience at one time or another with the Egyptian empire, in the Near East, North Africa, India, and the Far East. Without exception these civilizations retain a negative attitude towards the combination of 2nd and 4th efflorescence modes which enabled Egyptian civilization to generalize 3rd efflorescence society and to make civilization as we know it possible.

However, as the result of the social and political excesses of Egyptian civilization under which they suffered, these civilizations developed ego-negating and libido-abdicating orientations which were central in forming the particular configurations of these various civilizations. It is the absence of this negative attitude towards the combined 2nd and 4th efflorescence modes which made generalization of superego entrenchment possible which societies of the so-called 'under developed' world find so offensive about the so-called 'developed' world. From the point of view of these civilizations the excesses that resulted from a combination of 2nd and 4th efflorescence modes in Egyptian civilization were at least as important as its contributions to the welfare of mankind by way of universalized civilization. For these civilizations, an absence of ego-negating and libido-abdicating orientations bode ill for the future of the human race. Even China in the throes of Maoism appears to have undergone social and political excesses as a means to the end of achieving ego-negating and libido-abdicating orientations as a basis for collective, as opposed to individual, modes of thought and behaviour.

The best thinking of the so-called 'under developed' societies came to the conclusion, by way of both logic and experience, that unbridled ego-recognition and libido-aggression, either in individual or collective modes, would bring the human race to the brink of self-destruction. Many fallacies can be found in this argument from the point of view of societies in the 'developed' world, not the least of which is the unquestioned material progress of the 'developed' world, and in some instances at least the establishment of stable and non-aggressive collectives well entrenched in 4th efflorescence modes without pronounced ego-recognition and libido-aggression on either individual or collective levels. Yet from the point of view of the 'under developed' world, these are the arguments of collectives who lack the logical, moral or even linguistic precision to visualize the future consequences of their present actions.

The kind of anathema which 'under developed' and 'developed' societies feel for each other and which make up the cultural and racial conflicts that comprise the political and social geology of our times, is similar to the intrinsic anathema that Jung portrayed as existing between the extrovert and the introvert on an individual level, where the extrovert is represented by the 'developed' world in the 4th efflorescence and the 'introvert' is represented by the 'under developed' world in the late stages of the 3rd efflorescence. While introversion and extroversion may in part be the result of genetic inheritance, the historical experience of the individual, or the collective, plays a central role in the ultimate psychic orientation. Both of course may be neurotic, or even psychotic, but the pathological condition in individuals who possess these different modes is unrelated to the mode itself. Curative therapy, however, for the pathological conditions in these different modes necessarily demands differing approaches and results in differing solutions. For the problem of 'under developed' societies in late stages of the 3rd efflorescence, there has been an inadequate rational examination of past historical experiences which have led to present conditions and modes of behaviour and which have often, irrationally, been accepted as absolutes. For the problems of the 'developed' societies in the 4th efflorescence there has been an inadequate subjectification of an historical past that has touched them emotionally only in a tangential way, but from which all of their present cognitive conditions and modes have been ultimately gleaned. The absence of a sense of absolutism in the extroverted 'developed' societies is as much a problem as the presence of this sense in the introverted 'underdeveloped' societies. Only a synthesis of these two modes, the late stages of the 3rd efflorescence and the 4th, may allow the human species to leap the abyss of self-destruction into a new, 5th efflorescence mode.

Thus, in attempting to evaluate the 'master code' of the ancient Egyptian empire, or the Chinese *I-Ching*, which is the only extant version of this 'master code,' it is necessary to remember that it developed under special evolutionary and historical conditions and was specifically oriented for purposes in the context of these conditions. Yet while these historical and evolutionary conditions were special, they remain of intrinsic importance to the mode of historical evolution which has produced all the civilizations of the modern world, or civilization such as we know it and recognize it. To dismiss Egyptian civilization or its product, the 'master code,' as an accident of history is similar to dismissing the emergence of upright stance among mammals as an evolutionary 'accident.' Accidents though they may be, they are better described by the terms 'destiny' or 'fate' which implies that they are occurrances that, while defying logical explanation, supercede logical explanation. So intrinsic has this 'destiny' been to man's present condition, that we might say that it represents a logic beyond man's ability to discern, perhaps the thoughts and actions of the cosmic consciousness, or cosmic libido, or simply, God. The 'master code' represents the special conditions under which Egyptian civilization was born, and under whose long shadow we still continue to live in every part of the civilized world. While it may provide no direct solutions to the predicament of mankind today—again as much on the brink of self-extinction as he was in the peak period of 2nd efflorescence development—it may help us to better understand the exact circumstances of our predicament.

As the most significant record documenting the development of civilization that we possess, it remains the prototype of all the subsequent documents in our various civilizations. Thus there is the sense in our so-called 'developed' world that we are often travelling full circle, and in the process of progressive thinking rediscovering concepts and theories that we find to be ancient and already existing in the cultures of the so-called 'underdeveloped' world. Educated people in the 'developed' world read the texts of Taoism and Buddhism and are astounded at their profound psychological insights and at their general modernity. This sense of *deja vu* is not illusionary but real, as we find the psychological theories of Freud and Jung already clearly explicated in the 'master code' and even Marx's theories of historical evolution seem simplistic by comparison with the detailed tabulation of man's historical evolution in that document. Jung's insight that present conditions of each individual in terms of his psychic modes can be found in the collective unconscious are thus verified, as is Levi-Strauss' insight that the mind of man has a universal structure surpassing the artificial boundaries of nationality and race. So central are the special conditions in which Egyptian civilization developed and the 'master code' which tabulated these conditions to mankind's present predicament that they may, from the point of view of determining mankind's historical and evolutionary fate, both past, present, and future, be considered as an absolute. As another accident of history, or as a matter of fate, the all-important 'master code' was preserved for posterity by the Chinese, where out of anxiety, fear or in sheer recklessness, it was abandoned or destroyed by the other cultures whose common roots lay in this same document.

The development of civilization, like the development of any other organic system was born of trauma. The ravages of trauma in later stages of development can be cured only when this trauma can be clearly recalled and relived, if only in the imagination. The keys to the present trauma of civilization can only be found in the 'master code.' If we have the will and determination to utilize these keys, mankind in his civilization phase may yet be able to undo the complicated modes of maladaptive behaviour which threatens his survival as a species. Once the bewildering array of maladaptive behaviour which so characterizes our modern world is unravelled, mankind may yet have the chance, under the auspices of a benevolent fate, to proceed unhindered into a new 5th efflorescence mode of evolutionary development.

BIBLIOGRAPHY

Aberle, David F., *et al.*: The Incest Taboo and Mating Patterns of Animals, *American Anthropologist, Vol. 65*, pp. 253-65.

Allison, A.C.: Aspects of Polymorphism in Man, *Cold Spring Harbour Symposia in Quantitative Biology, Vol. 20*, pp. 239-55, 1955.

Ardrey, Robert: *The Territorial Imperative*, Antheum, N.Y., 1966.

Asimov, Isaac: *The Genetic Code*, New American Library, N.Y., 1962.

Baker, P.T.: The Biological Adaptation of Man to Hot Deserts, *American Naturalist, Vol. 92*, pp. 337-57, 1958.

Balazs, Etienne: *Chinese Civilization and Bureaucracy*, Yale U. Press, New Haven, 1967.

Barnicot, N.A.: Climatic Factors in the Evolution of Human Populations, *Cold Spring Harbour Symposia in Quantitative Biology, Vol. 24*, pp. 115-29, 1959.

Barnett, S.A., ed.: *A Century of Darwinism*, Harvard U. Press, Cambridge, Mass., 1959.

Barthes, Roland, S/Z: *Collection "Tel Quel,"* Editions du Seuil, Paris, 1970.

Bartholomew, George A., and Birdsell, Joseph B.: Ecology and the Protohominids, *American Anthropologist, Vol. 55*, pp. 481-98, 1953.

Benedict, Ruth: *Patterns of Culture*, New American Library, N.Y., 1959.

______: *Race, Science and Politics*, Modern Age Inc., N.Y., 1940.

Bergler, Edmund, M.D.: *Counterfeit-Sex: Homosexuality, Impotence, Frigidity*, Grove Press, N.Y., 1961.

Berne, Eric, M.D.: *Transactional Analysis in Psychotherapy*, Grove Press, N.Y., 1961.

Bettelheim, Bruno: *Symbolic Wounds, Puberty Rites and the Envious Male*, Collier Books, N.Y., 1962.

The Bhagavad Gita, tr. Juan Mascaro, Penguin, London, 1962.

Binford, Lewis R., and Binford, Sally R.: A Preliminary Analysis of Functional Variability in the Mousterian of Levallois Facies, *American Anthropologist, Vol. 68*, pp. 238-95, April, 1965.

Birdsell, J.B.: On Methods of Evolutionary Biology and Anthropology, *American Scientist, Vol. 45*, pp. 393-400, 1957.

Blalock, Hubert M., Jr.: *Theory Construction: From Verbal to Mathematical Formulations*, Prentice-Hall, N.J., 1969.

Blofeld, John: *The Book of Changes*, George Allen & Unwin, London, 1968.

Blyth, R.H. (the writings of): *Games Zen Masters Play*, New American Library, N.Y., 1976.

Boas, Franz: *Race, Language and Culture*, Free Press, N.Y., 1966.

Bohannan, Paul: *Africa and Africans*, Natural History Press, N.Y., 1964.

______: *Law and Warfare: Studies in the Anthropology of Conflict*, Natural History Press, N.Y., 1967.

______: *Social Anthropology*, Holt, Rinehart & Winston, N.Y., 1963.

Bohannan, Paul, and Middleton, John, eds.: *Kinship and Social Organization*, The Natural History Press, Garden City, N.Y., 1968.

Bonner, David M.: *Heredity: Foundations of Modern Biology*, Prentice-Hall, N.J., 1961.

Bordes, F.: Le Paléolithique Inférieure et Moyen de Jabrud (Syrie) et la Question due Pré-Aurignacien, *L'Anthropologie, Vol. 59*, Nos. 5-6, pp. 486-507, 1955.

Bordieu, Pierre: *Outline of a Theory of Practice*, Cambridge U. Press, London, 1977.

Bottomore, T.B., and Rubel, Maximilien: *Karl Marx: Selected Writings in Sociology and Social Philosophy*, Penguin, London, 1965.

______: *Classes in Modern Society*, George Allen & Unwin, London, 1965.

Boulding, Kenneth: *The Image*, University of Michigan Press, Ann Arbor, 1956.

Bourlière, Francois: *The Natural History of Mammals*, Alfrd A. Knopf, New York, 1956.

Brace, C. Loring: The Fate of the 'Classic' Neanderthals: A Consideration of Hominid Catastrophism, *Current Anthropology, Vol. 5*, pp. 3-43, 1964.

Brace C. Loring, and Montagu, M.F. Ashley: *Man's Evolution: An Introduction to Physical Anthropology*, Macmillan, N.Y., 1965.

Brinton, Crane, Crane: *The Anatomy of Revolution*, Vintage Books, N.Y., 1957.

Broom, Robert: The Age-Men, *Scientific American*, November, 1949.

Bruer, Alice: The Spearman and the Archer—An Essay on Selection in Body Build, *American Anthropologist, Vol. 61*, pp. 457-69, 1959.

Budge, E.A. Wallis: *The Gods of the Egyptians, Vols. I and II,* Dover Brooks, New York, 1904.

Bugelski, B.R., *An Introduction to the Principles of Psychology*, Bobbs-Merrill Co., Indianapolis, 1973.

Bulfinch, Thomas: *Mythology*, Modern Library, N.Y., n.d.

Bullock, Alan: *Hitler: A Study in Tyranny*, Harper & Brothers, N.Y., 1958.

Burgess, John Stewart: *The Guilds of Peking*, Ch'eng-Wen Publishing Co., Taipei, 1966.

Bushnell, G.H.S.: *Ancient Arts of the Americas*, Praeger, N.Y., 1965.

Campbell, Bernard G., ed.: *Humankind Emerging*, Little, Brown & Co., Boston, 1976.

Carpenter, C.R.: Behaviorism and Social Relations of the Howling Monkey, *Comparative Psychology Monographs*, May, 1934.

_______: *Naturalistic Behavior of Nonhuman Primates*, Pennsylvania State U. Press, 1964.

Cassirer, Ernest: *An Essay on Man*, Yale U. Press, New Haven, 1944.

_______: *The Myth of State*, Yale U. Press, New Haven, 1946.

Chomsky, Noam: *Language and Mind*, Harcourt, Brace, Jovanovich Inc., N.Y., 1968.

Clark, J. Desmond: *The Prehistory of Southern Africa*, Penguin, London, 1959.

Clark, W.E. Le Gros: *The Antecedents of Man: An Introduction to the Evolution of Primates*, Edinburgh U. Press, Edinburgh, 1959.

______: *The Fossil Evidence for Human Evolution: An Introduction to the Study of Paleoanthropology*, Second Edition, U. of Chicago Press, Chicago, 1964.

Clark, W.E. Le Gros, and Leakey, L.S.B.: The Miocene Hominoidea of East Africa, *Fossil Mammals of Africa, No. I, British Museum (Natural History)*, p. 1-115, 1951.

Coon, C.C.: *The Origin of Races*, Alfred A. Knopf, N.Y., 1962.

Coppens, Yves *et al.*, eds.: *Earliest Man and Environments in the Lake Rudolf Basin: Stratigraphy, Paleoecology and Evolution*, U. of Chicago Press, Chicago, 1976.

Dahrendorf, Ralf: *Class and Class Conflict in Industrial Society*, Stanford U. Press, 1959.

Darwin, C.G.: *The Next Million Years*, Doubleday, Garden City, N.Y., 1952.

Darwin, Charles: *The Descent of Man*, 1871. Revised edition, London, Merrill & Baker, London, 1874.

______: *On the Origin of Species*, 1859. Reprinted Everyman's Library, London, Dent & Sons, 1956.

Davies, D. Trevor: *Four Centuries of Witch-Beliefs*, Metheun, London, 1947.

Davis, Kingsley: *Human Society*, Macmillan, N.Y., 1949.

Day, M.H., and Napier, J.R.: A Hominid Toe Bone from Bed 1, Olduvai Gorge, Tanzania, *Nature, Vol. 211*, No. 5052, pp. 929-30, Aug. 27, 1966.

DeVore, Irven, ed.: *Primate Behavior: Field Studies of Monkeys and Apes*, Holt, Rinehart, & Winston, N.Y., 1965.

Dobzhansky, Th.: *Mankind Evolving: The Evolution of the Human Species*, Yale U. Press, New Haven, 1962.

Durkheim, Emile: *The Division of Labor in Society*, Free Press, N.Y., 1964.

______: *The Elementary Forms of the Religious Life*, Free Press, N.Y., 1965.

Eggan, Paul: *Social Organization of the Western Pueblos*, U. of Chicago Press, Chicago, 1950.

Einzig, Paul: *Primitive Money in its Ethnological, Historical and Economic Aspects*, Eyre & Spottiswoode, London, 1949.

Embree, Ainslie T., ed.: *The Hindu Tradition: Readings in Oriental Thought*, Vintage Books, N.Y., 1972.

Embree, John F.: *Suye Mura*, U. of Chicago Press, Chicago, 1939.

Etzioni, Amitai: *Studies in Social Change*, Holt, Rinehart & Winston, N.Y., 1966.

Evans-Pritchard, E.E.: *The Nuer: A Description of the Modes of Livlihood and Political Institutions of a Nilotic People*, Oxford U. Press, London, 1940.

______: *Witchcraft, Oracles and Magic Among the Azande*, Oxford: Clarendon Press, London, 1937.

Fan, K., ed.: *Mao Tse-Tung and Lin Piao: Post Revolutionary Writings*, Doubleday & Co., N.Y., 1972.

Feuer, Lewis, ed.: *Marx & Engles: Basic Writings on Politics and Philosophy*, Doubleday, N.Y., 1959.

Firth, Raymond: *We, the Tikopia: Kinship in Primitive Polynesia,* Beacon Press, Boston, 1963.

Forde, C. Daryll: *Habitat, Economy and Society*, Metheun, London, 1964.

Fortes, Meyer, and Evans-Pritchard, E.E.: *African Political Systems*, Oxford U. Press, London, 1940.

Foye, William O., ed.: *Principles of Medical Chemistry*, Lea & Febiger, Philadelphia, 1974.

Frazer, J.G.: *The Golden Bough*, Macmillan, London, 1922.

Freud, Sigmund: *Beyond the Pleasure Principle*, 1920. Reprinted, International Psychoanalytical Library, Lond, Hogarth Press, 1961.

______: *Civilization and Its Discontents*, 1930. Reprinted London, Hogarth Press, 1949.

______: *Collected Papers, Vol. I-IV*, Basic Books, N.Y., 1959.

______: *Outline of Psycho-Analysis*, 1940. Reprinted London, Hogarth Press, 1949.

______: *Totem and Taboo*, 1913. Reprinted London, Routledge & Kegan Paul, 1960.

Friedrich Engles: Eine Biographie, Autoren kollectiv: Gemkow, Heinrich (Leiter), Dietz Verlag, Berlin, 1970.

Gazin, C. Lewis: A Review of the Middle and Upper Eocene Primates of North America, *Smithsonian Miscellaneous Collections, Vol. 136*, No. 1, pp. 1-112, July, 1958.

Gilbert, William Harlen: *Peoples of India*, Smithsonian Institution, Washington, D.C. 1944.

Gleason, H.A.: *An Introduction to Descriptive Linguistics*, Holt, Rinehart, & Winston, N.Y., 1955.

Goffman, Erving: *Asylums: Essays on the Social Situation of Mental Patients and Other Inmates*, Doubleday & Co., Garden City, N.Y., 1961.

______: *Encounters: Two Studies in the Sociology of Interaction*, Bobbs-Merrill Co., Indianapolis, 1961.

______: *The Presentation of Self in Everyday Life*, Doubleday & Co., Garden City, N.Y., 1959.

______: *Stigma: Notes on the Management of Spoiled Identity*, Prentice-Hall, N.J., 1963.

The Golden Lotus, (a translation of the Chinese novel: *Ch'ing Ping Mei*), *Vol. I-IV*, tr. by Clement Egerton, Routledge & Kegan Paul, London, 1957.

Goodall, Jane van Lawick: Feeding Behaviour of Wild Chimpanzees, *Symposium Zoological Society, London, Vol. 10*, pp. 37-47, August, 1963.

______: My Life Among Wild Chimpanzees, *National Geographic, Vol. 124*, pp. 272-308, August, 1963.

______: New Discoveries Among Wild Chimpanzees, *National Geographic, Vol. 128*, pp. 802-831, Dec., 1965.

Goody, Jack, ed.: *The Development Cycle in Domestic Groups*, Cambridge U. Press, London, 1958.

Greenberg, Joseph H.: *Essays in Linguistics*, U. of Chicago Press, Chicago, 1957.

Grossman, Carl M., M.D., and Grossman, Sylvia: *The Wild Analyst: The Life and Work of George Groddeck*, A Delta Book, Dell Publishing Co., N.Y., 1965.

Haldane, J.B.S.: *The Causes of Evolution*, Harper & Row, N.Y., 1932.

Hamburger, Michael: *Reason and Energy: Studies in German Literature*, Grove, N.Y., 1957.

Hamilton, Edith: *Mythology: Timeless Tales of Gods and Heroes*, New American Library, N.Y., 1969.

Hanson, Earl D.: *Animal Diversity: Foundation of Modern Biology Series*, Prentice-Hall, N.J., 1961.

Harmon, Harry H.: *Modern Factor Analysis*, U. of Chicago Press, 1967.

Harris, Marvin: *Cannibals and Kings: The Origins of Culture*, Vintage Books, N.Y., 1978.

Harrison, G.A., Weiner, J.S., Tanner, J.M., and Barnicott, N.A.: *Human Biology: An Introduction to Human Evolution, Variation and Growth*, Oxford U. Press, N.Y., 1964.

Harrison, Jane: *Prolegomena to the Study of Greek Religion*, Cambridge U. Press, 1963.

Harrison, R.J.: *Man the Peculiar Animal*, Penguin, London, 1958.

Hayaishi, Osamu, and Asada, Kozi, eds.: *Biochemical and Medical Aspects of Active Oxygen: Papers Presented at a Symposium Held in Kyoto, Japan, Nov. 29, 1976*, University Park Press, Baltimore, 1977.

Henle, Paul: *Language, Thought and Culture*, U. of Michigan Press, Ann Arbor, 1968.

Herskovits, Melville J.: *Economic Anthropology: The Economic Life of Primitive Peoples*, W.W. Norton, N.Y., 1952.

Hinton, Harold C.: *China's Turbulent Quest: An Analysis of China's Foreign Relations Since 1949*, Indiana U. Press, Indianapolis, 1972.

———: *An Introduction to Chinese Politics*, Praeger, N.Y., 1975.

Hjelmslev, L.: *Prolegomena to a Theory of Language, University of Indiana Publications in Anthropology and Linguistics, No. 7*, Waverly Press, Baltimore, 1953.

Hockett, Charles F.: Chinese vs. English: An Exploration of the Whorfian Hypothesis, In *Language and Culture* (H. Hoijer, ed.), *American Anthropological Association Memoir*, No. 79, 1954.

Hogart, A.M.: *Caste: A Comparative Study*, Metheun, London, 1950.

Hoebel, E.A.: *The Law of Primitive Man*, Harvard U. Press, Cambridge, 1954.

———: *Man in the Primitive World: An Introduction to Anthropology*, McGraw-Hill, N.Y., 1949.

Hoebel, E.A., and Wallace, E.: *The Comanches: Lords of the South Plains*, U. of Oklahoma Press, Norman, 1952.

Hole, Frank, and Heizer, Robert F.: *An Introduction to Prehistoric Archaeology*, Holt, Rinehart, and Winston, N.Y., 1966.

Holton, Gerald: *The Scientific Imagination: Case Studies*, Cambridge U. Press, London, 1978.

The Holy Bible: Containing The Old and New Testaments, Commonly Known as the Authorized (King James) Version, The National Publishing Co., 1961.

Homans, G.C., and Schneider, D.M.: *Marriage, Authority and Final Causes*, Free Press, N.Y., 1955.

Hooke, S.H.: *Middle Eastern Mythology*, Penguin Books, London, 1978.

Hooten, Ernest Albert: *Up from the Ape*, Macmillan, N.Y., 1946.

Horney, Karen, M.D.: *Neurosis and Human Growth: The Struggle Towards Self-Realization*, W.W. Norton, N.Y., 1950.

Howell, F. Clark: The Evolutionary Significance of Variation and Varieties of *Neanderthal* Man, *Quarterly Review of Biology, Vol. 32*, pp. 330-347, 1957.

______: The Villafrancian and Human Origins, *Science, Vol. 130*, pp. 831-844, 1959.

______: European and Northwest African Middle Pleistocene Hominids, *Current Anthropology, Vol. 1,* pp. 195-232, 1960.

Howells, W.W.: *Mankind in the Making*, Doubleday, N.Y., 1959.

Hsu, Francis L.K.: *Clan, Caste and Club*, Princeton: Van Nostrand, 1963.

______: *Under the Ancestor's Shadow*, Doubleday, N.Y., 1967.

Hutton, O.H.: *Caste in India*, Cambridge U. Press, London, 1946.

Huxley, Julian S.: *Evolution: The Modern Synthesis*, Harper & Row, N.Y., 1942.

Huxley, T.H., and Huxley, J.S.: *Touchstone for Ethics*, Harper & Row, N.Y., 1947.

Hymes, Dell: *Language in Culture and Society: A Reader in Linguistics and Anthropology*, Harper & Row, N.Y., 1964.

Isaac, Glynn, and Leakey, Richard E.F., eds.: *Human Ancestors: Readings from Scientific American*, W.H. Freeman & Company, San Francisco, 1979.

Jalee, Pierre: *The Third World in World Economy*, Monthly Review Press, N.Y., 1969.

Jung, C.G.: *The Collected Works*, Bollingen Series XX, Princeton U. Press, 1976.

Kalmus, Hans: *Genetics*, West Drayton: Pelican Books, 1948.

Kardiner, Abram, and Preble, Edward: *They Studied Man*, World Publishing Co., 1963.

Keeley, Lawrence H., and Newcomer, Mark H.: Microwear Analysis of Experimental Flint Tools: A Test Case, *Journal of Archaeological Science, Vol. 4,* No. 1, pp. 29-62, March, 1977.

Kennan, George F.: *Russia and the West*, New American Library, 1961.

Klemm, Frederick: *A History of Western Technology*, M.I.T. Press, Cambridge, Mass., 1975.

Kluckholm, Clyde, and Murray, H. (eds.): *Personality in Nature, Society and Culture*, Alfred A. Knopf, N.Y., 1949.

Koestler, Arthur: *The Ghost in the Machine*, Pan Books, London, 1975.

Korn, Noel, and Thompson, F. (eds.): *Human Evolution: Readings in Physical Anthropology*, Holt, Rinehart and Winston, N.Y., 1967.

Kramer, Samuel Noah: *History Begins at Sumer*, Doubleday, Garden City, N.Y., 1959.

Kroeber, A.L.: *Anthropology: Cultural Patterns and Process*, Harcourt, Brace & World, N.Y., 1963.

———: *An Anthropologist Looks at History*, U. of California Press, Berkeley, 1963.

Kroeber, A.L., and Kluckhohm, Clyde: *Culture: A Critical Review of Concepts and Definitions*, Vintage Books, N.Y., 1963.

Kulp, Daniel Harrison: *Country Life in South China*, Bureau of Publications, Columbia U., N.Y., 1925.

Lacan, Jacques: *De la Psychose Paranöis dans ses rapports avec la personalité*, Editions du Seuil, Paris, 1973.

Lancaster, Jane B.: *Primate Behavior and the Emergence of Human Culture*, Holt, Rinehart and Winston, 1975.

Laing, R.D.: *The Divided Self: An Existential Study in Sanity and Madness*, Penguin, London, 1973.

Leach, Edmund: *Lévi-Strauss*, Fontana, London, 1963.

———: *Political Systems of Highland Burma*, Beacon Press, Boston, 1954.

———: *Rethinking Anthropology, London School of Economics Monographs on Anthropology, No. 22*, U. of London, Athelone Press, 1966.

Leakey, L.S.B.: A New Fossil Skull from Olduvai, *Nature, Vol. 184*, pp. 491-493, 1959.

______: *Adam's Ancestors*, Metheun, London, 1953.

Leakey, May, and Leakey, Richard E.F.: *Koobi for a Research Project, Vol. 1: The Fossil Hominids and an Introduction to Their Context*, 1968-1974, Oxford U. Press, London, 1978.

Leakey, Richard E., and Lewin, Roger: *People of the Lake: Mankind and Its Beginnings*, Doubleday, Garden City, N.Y., 1978.

Lee, Dorothy: *Freedom and Culture*, Prentice-Hall, N.J., 1959.

Legge, James: tr., *The I-Ching: The Book of Changes*, Clarendon Press, 1899, reprinted by Dover, N.Y., 1963.

______: *The Works of Mencius*, 1894, reprinted by Dover, N.Y., 1970.

Lehninger, Albert L.: *Biochemistry*, Worth Publishers, Inc., N.Y., 1978.

Lenin, V.I.: *Left Wing Communism: An Infantile Disorder*, Foreign Language Press, Peking, 1970.

______: *Selected Works*, Progress Publishers, Moscow, 1971.

Leslie, Charles: *Anthropology of Folk Religion*, Vintage Books, 1960.

Lévi-Strauss, Claude: *The Elementary Structures of Kinship*, Eyre & Spottiswode, Ltd., London, 1969.

______: *The Raw and the Cooked: Introduction to a Science of Mythology*, Harper & Row, N.Y., 1964.

______: *The Savage Mind*, Weidenfeld and Nicolson, London, 1966.

______: *The Scope of Anthropology*, Jonathan Cape, London, 1968.

______: *Triste Tropiques: An Anthropological Study of Primitive Societies in Brazil*, Atheneum, N.Y., 1965.

Lévy-Bruhl, Lucien: *The Soul of the Primitive*, George Allen & Unwin, London, 1965.

Lewin, Kurt: *Selected Papers: A Dynamic Theory of Personality*, McGraw-Hill, N.Y., 1935.

Lewis, Oscar: *The Children of Sanchez: An Autobiography of a Mexican Family*, Vintage Books, N.Y., 1961.

______: *Village Life in Northern India*, Vintage Books, N.Y., 1965.

Litwack, Gerald, ed.: *Biochemical Aspects of Hormones*, Academic Press, N.Y., 1979.

Lorenz, Konrad Z.: The Comparative Method in Studying Innate Behavior Patterns, *Symposia Society Exp. Biology, Vol. 4*, pp. 221-268, 1950.

______: *King Solomon's Ring*, Crowell Publishers, N.Y., 1952.

Malinowski, Bronislaw: *The Dynamics of Culture Change: An Inquiry into Race Relations in Africa*, Yale U. Press, New Haven, 1961.

______: *The Father in Primitive Psychology*, W.W. Norton, N.Y., 1966.

______: *The Family Among the Australian Aborigines*, Schocken Books, N.Y., 1963.

______: *Magic, Science, Religion and other Essays*, Free Press, N.Y., 1948.

Mallery, Arlington, and Harrison, Mary Roberts: *The Rediscovery of Lost America: The Story of the Pre-Columbian Iron Age in America*, Dutton, N.Y., 1979.

Mann, Dr. Felix: *Acupuncture: Cure of Many Diseases*, Pan Books Ltd., London, 1971.

Marx, Karl: *Capital, Vols. I and II* (ed. by Frederik Engels), International Publishers, N.Y., 1971.

______: *The Poverty of Philosophy*, Progress Publishers, Moscow, 1975.

______: *Selected Writings* (eds. T.B. Bottomore and Maximillien Rubel), Penguin, London, 1965.

May, Rollo, *et al.*, eds.: *Existence: A New Dimension in Psychiatry and Psychology*, Basic Books, N.Y., 1959.

Mayr, Ernst: *Animal Species and Evolution*, Belknap Press, Harvard U. Press, Cambridge, Mass. 1963.

Mead, George H.: *Mind, Self and Society*, U. of Chicago Press, Chicago, 1967.

Mead, Margaret: *Male and Female: A Study of the Sexes in a Changing World*, New American Library, N.Y., 1962.

Mead, Margaret, and Bunzell, Ruth L., eds.: *The Golden Age of American Anthropology*, Braziller, N.Y., 1960.

Mead, Margaret, and Wolfenstein, Martha, eds.: *Childhood in Contemporary Cultures*, U. of Chicago Press, Chicago, 1955.

Mencius: *Works*, tr. by D.C. Lau, Penguin Books, London, 1970.

Merrill, D.J.: *Evolution and Genetics: The Modern Theory of Evolution*, Holt, Rinehart, and Winston, N.Y., 1962.

Montagu, M.F. Ashley: *The Biosocial Nature of Man*, Grove Press, N.Y., 1956.

______: *Culture and the Evolution of Man*, Oxford U. Press, London, 1962.

______: *The Humanization of Man*, World Publishing, Cleveland and N.Y., 1962.

Montagu, M.F. Ashley, ed: *The Concept of Race*, Free Press, N.Y., 1964.

Morris, Eleanor B.: (*see* Wu, Eleanor B. Morris).

Mosca, Gaetano: *The Ruling Class* (ed. by Arthur Livingston), McGraw-Hill, N.Y., 1965.

Mumford, Lewis: *The Myth of the Machine: The Pentagon of Power*, Harcourt, Brace, Jovanovich, N.Y., 1970.

Napier, J.R.: The Foot and the Shoe, *Physiotherapy, Vol. 43*, No. 3, pp. 65-74, March, 1957.

______: The Prehensile Movement of the Human Hand, *The Journal of Bone and Joint Surgery, Vol. 38-B*, No. 4, pp. 902-913, Nov., 1956.

______: Prehensibility and Opposability in the Hands of Primates, *Symposia of the Zoological Society of London*, No. 5, pp. 115-132, Aug., 1961.

Newman, Philip L.: *Knowing the Gururumba*, Holt, Rinehart and Winston, N.Y., 1965.

Nietzsche, Friedrick: *Der Antichrist; Ecco Homo; Dionysos-Dithyrambus*. William Goldmann Verlag, München, 1979.

______: *Also Sprach Zarathustra*, Philipp Reclam, Jun. Stuttgart, 1978.

Oakley, Kenneth P.: *Man the Tool-Maker*, British Museum of Natural History, London, 1950.

Oliver, Douglas L.: *Invitation to Anthropology: A Guide to Basic Concepts*, The Natural History Press, Garden City, N.Y., 1964.

Ostrander, Shiela, and Shroeder, Lynn: *Psychic Discoveries Behind the Iron Curtain*, Bantem, N.Y., 1971.

Palmer, R.R., and Colton, Joel: *A History of the Modern World*, Alfred A. Knopf, N.Y., 1960.

Penfield, Wilder, and Rasmusse, Theodore: *Cerebral Cortex of Man*, Macmillan, N.Y., 1950.

Piaget, Jean: *The Language and Thought of the Child*, Routledge and Kegan Paul, London, 1965.

Pilbeam, David: Newly Recognized Mandible of Ramapithecus, *Nature, Vol. 222*, No. 5198, pp. 1093-4, June 14, 1969.

Pinson, Koppel S.: *Modern Germany: Its History and Civilization*, Macmillan, N.Y., 1959.

Pipes, Richard: *Russia Under the Old Regime*, Penguin, London, 1974.

Popp, Fritz Albert, Becker, Günter *et al.*, eds.: *Electromagnetic Bio-Information: Proceedings of the Symposium, Marburg, Sept. 5, 1977*, Urban & Swarzenberg, München, Baltimore, 1979.

Radcliffe-Brown, A.R.: *Structure and Function in Primitive Society*, Free Press, N.Y., 1952.

Rapaport, Anatol: *Two-Person Game Theory and the Essential Ideas*, U. of Michigan Press, Ann Arbor, 1966.

Redfield, Robert: *The Little Community: Peasant Society and Culture*, U. of Chicago Press, Chicago, 1967.

Rees, Alwyn, and Rees, Brinley: *Celtic Heritage: Ancient Tradition in Ireland and Wales*, Thames and Hudson, London, 1975.

Reich, Wilhelm: *The Mass Psychology of Fascism*, Simon & Schuster, N.Y., 1970.

Rensch, B.: *Evolution Above the Species Level*, Columbia U. Press, N.Y., 1960.

Richards, P.W.: *The Tropical Rain Forest*, Cambridge U. Press, London, 1957.

Roberts, D.F.: Body Weight, Race and Climate, *American Journal of Physical Anthropology, Vol. 11*, pp. 533-558, 1953.

Ruitenbeek: *Psychoanalysis and Social Sciences*, Dutton, N.Y., 1962.

Sapir, Edward: *Language: An Introduction to the Study of Speech*, Harcourt, Brace and World, N.Y., 1949.

Schaller, George B.: *The Year of the Gorilla*, U. of Chicago Press, Chicago, 1964.

_______: *The Mountain Gorilla: Ecology and Behavior*, U. of Chicago Press, Chicago, 1963.

Schmit-Nielsen, K.: *Animal Physiology*, Prentice-Hall, N.Y., 1960.

Schram, Stuart R.: *The Political Thought of Mao Tse-Tung*, Frederick A. Praeger, N.Y., 1969.

Schultz, A.H.: The Specialization of Man and His Place Among Catarrhine Primates, *Cold Spring Harbor Symposia on Quantitative Biology, Vol. 15*, pp. 37-53, 1950.

Semenov, S.A.: tr. M.W. Thompson: *Prehistoric Tehnology: An Experimental Study of the Oldest Tools and Artifacts from Traces of Manufacture and Wear*, Barnes & Noble, N.Y., 1964.

Sen, R.M.: *Hinduism*, Penguin, London, 1978.

Sheppard, P.M.: Blood Groups and Natural Selection, *British Medical Bulletin, Vol. 15*, pp. 134-9, 1959.

Shirer, William L.: *The Rise and Fall of the Third Reich: A History of Nazi Germany*, Simon & Schuster, N.Y., 1959.

Sih, Paul K.T., ed.: *The Strenuous Decade: China's Nation Building Efforts: 1927-1937*, St. John's U. Press, N.Y., 1976.

Simons, Elwyn L.: A Critical Reappraisal of Tertiary Primates, in *Evolutionary and Genetic Biology of Primates, Vol. I.*, Academic Press, N.Y., 1963.

———: Some Fallacies in the Study of Hominid Phylogeny, *Science, Vol. 141*, No. 3584, pp. 879-889, Sept., 1963.

———: *Primate Evolution: An Introduction to Man's Place in Nature*, Macmillan, N.Y., 1972.

Simmel, George *et al.*, eds.: *Essays on Sociology, Philosophy and Aesthetics*, Harper & Row, N.Y., 1965.

Simmel, George: *Sociology* (tr. and ed., Kurt H. Wolff), Free Presss, N.Y., 1950.

Simpson, George Gaylord: The Biological Nature of Man, *Science, Vol. 152*, pp. 472-478, 1966.

———: *Life of the Past*, Yale U. Press, New Haven, 1961.

Simpson, William Kelly: *The Literature of Ancient Egypt: An Anthology of Stories, Instruction and Poetry*, U. Yale Press, New Haven, 1976.

Singer, Milton: *When a Great Tradition Modernizes: An Anthropological Approach to Indian Civilization*, Praeger, 1972.

Smith, Alfred G., ed.: *Communication and Culture: Readings in the Codes of Human Interaction*, Holt, Rinehart, and Winston, N.Y., 1966.

Solomon, Richard H.: *Mao's Revolution and the Chinese Political Culture*, U. of California Press, Berkeley, 1971.

Soustelle, Jacques: *The Daily Life of the Aztecs*, Penguin, London, 1964.

Southwick, C.H., ed.: *Primate Social Behavior*, Van Nostrand, Princeton, N.J., 1963.

Sprott, W.J.H.: *Human Groups*, Penguin, London, 1966.

Spuhler, J.N., ed.: *The Evolution of Man's Capacity for Culture*, Wayne State U. Press, Indiana, 1959.

Strayer, Joseph R., and Munro, Dana C.: *The Middle Ages: 395-1500*, Appleton, Century, Crofts, N.Y., 1959.

Sung, Z.D.: *The Text of the Yi King (and Its Appendixes)* (Chinese Original with English Translation), Culture Book Co., Taipei, n.d.

Sun Tzu: *The Art of War: The Oldest Military Treatise in the World*, The Grand Cultural Service Co., Kowloon, Hong Kong, n.d.

Swami, Probhavanda, and Manchester, Frederick: *The Upanishads, Breath of the Eternal*, The Principal Texts Selected and Translated from the Original, New American Library, N.Y., 1957.

Talbot, Michael: *Mysticism and the New Physics*, Bantam, N.Y., 1980.

Tao Te Ching: Lao Tsu, tr. Gia Fu Feng and Jane English, Viking Books, N.Y., 1972.

Tax, Sol, ed.: *Evolution after Darwin; The University of Chicago Centennial, Vol. I. The Evolution of Life. Its Origin, History and Future, Vol. II. The Evolution of Man, Mind, Culture and Society*, U. of Chicago Press, Chicago, 1960.

______: *The Evolution of Man*, U. of Chicago Press, Chicago, 1960.

Teilhard de Chardin, P., *The Phenomenon of Man*, Harper & Row, N.Y., 1959.

Thorpe, W.H.: *Learning and Instincts in Animals*, Harvard U. Press, Cambridge, Mass., 1963.

Thucydides: *The Peloponnesian War*, Modern Library, N.Y., 1951.

Tinbergen, N.: *The Herring Gull's World*, Basic Books, N.Y., 1961.

______: *Social Behaviour in Animals*, John Wiley & Sons, N.Y., 1953.

______: *The Study of Instinct*, Oxford: Clarendon Press, London, 1951.

Toffler, Alvin: *Future Shock*, Random House, N.Y., 1971.

Trotsky, Leon: *Literature and Revolution*, U. of Michigan Press, Ann Arbor, 1960.

Tsao, Hsueh Chin: *Dream of the Red Chamber*, tr. by Chi-Chen Wang, Twayne Publishers, N.Y., 1958.

Tylor, Sir Edward Burnett: *The Origins of Culture*, Harper & Row, 1958.

Veblen, Thorsten: *The Higher Learning*, Sagamore Press, N.Y., 1957.

———: *Theory of the Leisure Class*, Mentor, New American Library, N.Y., 1953.

Velikovsky, Immanuel: *Ages in Chaos: A Reconstruction of Ancient History from the Exodus to King Akhnaton*, Doubleday, N.Y., 1952.

Von Neumann, John, and Morgenstern, Oskar: *Theory of Games and Economic Behavior*, John Wiley & Sons, N.Y., 1964.

Vygotsky, Lev S.: *Mind in Society*, Harvard U. Press, Cambridge, Mass., 1978.

———: *Thought and Language*, tr. Eugena Hangmann, Gertrude Vakar, M.I.T. Press, Cambridge, Mass., 1962.

Waley, Arthur: *Three Ways of Thought in Ancient China*, Macmillan, London, 1939.

———: *The Way and Its Power: A Study of the Tao Te Ching and It's Place in Chinese Thought*, George Allen & Unwin, London, 1965.

Waite, E.A.: *The Holy Kabbalah*, University Books Inc., Secaucus, New Jersey, 1975.

Wann, T.W., ed.: *Behaviorism and Phenomenology: Contrasting Bases for Modern Psychology, A Rice University Semi-Centennial Publication*, U. of Chicago Press, Chicago, 1964.

Washburn, S.L., ed.: *Classification and Human Evolution*, Aldine Publishing Co., Chicago, 1963.

———: *The Social Life of Early Man*, Wenner-Gren Foundation For Anthropological Research, N.Y., 1961.

Washburn, S.L., and DeVore, I.: The Social Life of Baboons, *Scientific American, Vol. 204*, pp. 62-71, 1961.

Watson, J.B.: *Behaviorism*, W.W. Norton, N.Y., 1924.

Watts, Alan: *Psychotherapy, East & West*, Pantheon Books, N.Y., 1961.

———: *The Supreme Identity, An Essay on Oriental Metaphysic and the Christian Religion*, Vintage Books, N.Y., 1972.

———: *Tao: The Watercourse Way*, Pantheon Books, N.Y., 1975.

———: *The Way of Zen*, Vintage Books, N.Y., 1957.

Weber, Max: *The Protestant Ethic and the Spirit of Capitalism*, Scribners, N.Y., 1930.

———: *The Religion of China*, Free Press, N.Y., 1964.

______: *The Sociology of Religion*, Beacon Press, Boston, 1964.
______: *The Theory of Social and Economic Organization*, Oxford U. Press, London, 1942.
______: *On Charisma and Institution Building*, U. of Chicago Press, Chicago, 1968.
Wei Tat: *An Exposition of the I-Ching or Book of Changes*, Institute of Cultural Studies, Yang Ming Shan Villa, Taipei, Taiwan, 1970.
Weiner, J.S.: Physical Anthropology . . . An Appraisal, *American Scientist, Vol. 45*, pp. 79-87, 1957.
Werner, E.T.C.: *A Dictionary of Chinese Mythology*, The Julian Press Inc., N.Y., 1969.
White, Leslie: *The Evolution of Culture: The Development of Civilization to the Fall of Rome*, McGraw-Hill, N.Y., 1959.
Whiteheard, Alfred N.: *Science and the Modern World, Lowell Lectures, 1925*, New American Library, N.Y., 1959.
Whorf, Benjamin Lee: *Language, Thought and Reality*, ed., John Carroll, M.I.T. Press, Cambridge, Mass., 1956.
Wilhelm, Helmut: *Change: Eight Lectures on the I-Ching*, The Bollingen Library, Harper & Row, N.Y., 1960.
Wilson, Edward O.: *Sociobiology*, Belknap Press, Harvard U. Press, Cambridge, Mass., 1980.
Wittfogel, Karl A.: *Oriental Despotism: A Comparative Study of Total Power*, Yale U. Press, New Haven, 1957.
Wittgenstein, Ludwig: *Tractatus Logico-Philosophicus*, Routledge and Kegan Paul, London, 1961.
______: *Notebooks, 1914-1916*, Harper & Row, N.Y., 1969.
______: *Remarks on the Foundations of Mathematics*, M.I.T. Press, Cambridge, Mass, 1975.
Wolf, Eric R.: *Peasants: Foundations of Modern Anthropology Series*, Prentice-Hall, N.J., 1966.
Wright, Arthur F., ed.: *Studies in Chinese Thought*, U. of Chicago Press, 1953.
Wright, Sewall: Evolution in Mendalian Populations, *Genetics, Vol. 16*, pp. 97-159, 1939.
Wu, Eleanor B. Morris: *Functions and Models of Modern Biochemistry in I-Ching*, Cheng Chung Book Co., Taipei, Taiwan, 1975.

———: Modern Philosophy, East and West, *I-Ching Quarterly*, Winter, Taipei, Taiwan, 1980.

———: *Information Puzzles and Astronomical Predictions in I-Ching*, Cheng Chung Book Co., Taipei, Taiwan, 1975.

———: Maoist *Contradictions* in Canada, *The Journal of Social and Political Studies, Vol. 3*, No. 2, Summer, Washington, D.C., 1978.

———: Introduction to Structural Biochemistry Based on Structural Principles Derived from the Chinese *Book of Changes*, *Chinese Culture: A Quarterly Review, Vol. XXII*, No. 2, June, 1981.

Wynne-Edwards, V.C.: *Animal Dispersion in Relation to Social Behavior*, Hafner Press, N.Y., 1962.

———: Population Control in Animals, *Scientific American, Vol. 211*, pp. 68-74, Aug., 1964.

———: Self-Regulatory Systems in Populations of Animals, *Science, Vol. 147*, pp. 1543-48, 1965.

Xuequin, Cao: *The Story of the Stone, A Chinese Novel in Five Volumes*, tr. by David Hawkes, Penguin, London, 1973.

Yang, C.K.: *Religion In Chinese Society*, U. of California Press, Berkeley, 1967.

Yang, Martin M.C.: *Chinese Social Structure: A Historical Study*, Eurasia Book Co., Taipei, Taiwan, 1969.

The Yellow Emperor's Classic of Internal Medicine: tr. Ilza Veith, U. of California Press, Berkeley, 1972.

Yudkin, Michael, and Offord, Robin: *A Guidebook to Biochemistry: A New Edition of a Guidebook to Biochemistry by K. Harrison*, Cambridge U. Press, London, 1971.

Yuho, Yokoi: *Zen Master Dogen: An Introduction with Selected Writings*, John Weatherhill, Inc., N.Y., 1976.

Zuckerman, S.: *The Social Life of Monkeys and Apes*, Routledge and Kegan Paul, London, 1932.

INDEX

—D—

—E—

—J—

—K—

—L—

—M—

—P—